AIR POWER

An Overview of Roles

Brassey's Air Power:
Aircraft, Weapons Systems and Technology Series

VOLUME 1

Brassey's Air Power:

Aircraft, Weapons Systems and Technology Series

General Editor: AIR VICE MARSHAL R. A. MASON, CBE, MA, RAF

This new series, consisting of eleven volumes, is aimed at the international officer cadet or junior officer level and is appropriate to the student, young professional and interested amateur who seek a sound basic knowledge of the technology of air forces. Each volume, written by an acknowledged expert, identifies the responsibilities and technical requirements of its subject and illustrates it with British, American, Russian, major European and Third World examples drawn from recent history and current events. The series is similar in approach and presentation to the highly successful Brassey's Battlefield Weapons Systems and Technology Series, and each volume, excluding the first, has self-test questions and answers.

Volume 1. Air Power: An Overview of Roles
　　　　AIR VICE MARSHAL R. A. MASON, CBE, MA, RAF

Volume 2. Air-to-Ground Operations
　　　　AIR VICE MARSHAL J. R. WALKER, CBE, AFC, RAF

Titles of Related Interest

R. A. MASON
War in the Third Dimension: Essays in Contemporary Air Power

J. GODDEN
Harrier: Ski Jump to Victory

B. MYLES
Jump Jet, 2nd Edition

P. G. HARRISON *et al.*
Military Helicopters

AIR POWER
An Overview of Roles

Air Vice Marshal R. A. Mason, CBE, MA, RAF

BRASSEY'S DEFENCE PUBLISHERS

(a member of the Pergamon Group)

LONDON · OXFORD · WASHINGTON · NEW YORK
BEIJING · FRANKFURT · SÃO PAULO · SYDNEY · TOKYO · TORONTO

UK (Editorial)	Brassey's Defence Publishers, 24 Gray's Inn Road, London WC1X 8HR
(Orders)	Brassey's Defence Publishers, Headington Hill Hall, Oxford OX3 0BW, England
USA (Editorial)	Pergamon-Brassey's International Defense Publishers, 1340 Old Chain Bridge Road, McLean, Virginia 22101, USA
(Orders)	Pergamon Press, Maxwell House, Fairview Park, Elmsford, New York 10523, USA
PEOPLE'S REPUBLIC OF CHINA	Pergamon Press, Qianmen Hotel, Beijing, People's Republic of China
FEDERAL REPUBLIC OF GERMANY	Pergamon Press, Hammerweg 6, D-6242 Kronberg, Federal Republic of Germany
BRAZIL	Pergamon Editora, Rua Eça de Queiros, 346, CEP 04011, São Paulo, Brazil
AUSTRALIA	Pergamon-Brassey's Defence Publishers, P.O. Box 544, Potts Point, NSW 2011, Australia
JAPAN	Pergamon Press, 8th Floor, Matsuoka Central Building, 1–7–1 Nishishinjuku, Shinjuku-ku, Tokyo 160, Japan
CANADA	Pergamon Press Canada, Suite 104, 150 Consumers Road, Willowdale, Ontario M2J 1P9, Canada

First edition 1987

Library of Congress Cataloging in Publication Data
Air power.
Includes index.
Contents: v. 1. Air power, an overview of roles/
R.A. Mason.
1. Air power. 2. Aeronautics, Military. 3. Air warfare.
I. Mason, R. A.
UG630.A3827 1986   358.4'03   86-21181

British Library Cataloguing in Publication Data
Mason, R.A.
Air power: an overview of roles.—(Air power; v.1)
1. Air power
I. Title II. Series
358.4   UG630

ISBN 0-08-031195-4 (Hardcover)
ISBN 0-08-031194-6 (Flexicover)

The front cover photograph shows a USAF F-16 equipped with two Hughes Aircraft Company advanced medium-range air-to-air missile (AMRAAM) prototype models banking during captive carry evaluation flight near Eglin Air Force Base, Florida (Hughes Aircraft Company).

Printed in Great Britain by A. Wheaton & Co. Ltd., Exeter

Preface

This volume sets out to give an introduction to and overview of the interaction of modern technology and air power, with particular emphasis on developments which seem likely to become more influential in the foreseeable future.

I am indebted to aerospace companies worldwide for technical data and illustrations of their products, but reference to any one aircraft or weapon system does not indicate any endorsement of its proclaimed effectiveness. Specific references and illustrations have been chosen as far as possible to be representative of major developments in each area of activity. No critical analyses or comparisons have been made, and it is fully appreciated that many factors other than integral technology affect weapon performance in combat, not all of them necessarily present in development trials. Nevertheless, the main thrusts of technological impact on air power in the last two decades are discernible and will be explored and analysed in detail in subsequent volumes in this series.

My particular thanks go to Chris Hobson, head librarian at the RAF Staff College, for his constant supply of supplementary source material, and to Ann Pearson for her unfailing ability to produce orderly manuscripts from illegible scrawl without ever losing her good humour.

Finally, the representation of facts and opinions in the following chapters are solely the author's own responsibility and imply no endorsement by the British Ministry of Defence or by any other agency.

R.A.M.

About the Author

Air Vice Marshal R. A. Mason, CBE, MA, RAF, is Air Secretary and Director General of Personnel for the Royal Air Force which he joined after reading history at St Andrews University. He was awarded an MA with Distinction in War Studies at King's College, London in 1967 and since then, in association with his RAF duties, he has written and lectured internationally on defence matters with particular reference to air power and the Soviet Air Forces. Previous publications include *Air Power in the Next Generation* (ed.), *Readings in Air Power, Air Power in the Nuclear Age* (with M. J. Armitage), *The Royal Air Force Today and Tomorrow, British Air Power in the 1980s, The Soviet Air Forces* and *War in the Third Dimension* (ed.) published by Brassey's.

Contents

List of Figures

List of Plates

1

Military Aircraft

Air power is about the exploitation of the third dimension above the land and sea by man, but not necessarily 'with' him. The advent of surface-to-surface and ground-to-air missiles, as well as the use of unmanned vehicles in the sky and space above it, offer ways to project military force above the earth's surface without dependence on manned aircraft. But for the foreseeable future air power will continued to be largely the responsibility of the aeroplane and the helicopter. Any study of the interaction of air power and technology must therefore begin with a survey of military aircraft themselves. Even that definition can be ambiguous, because at one end of the spectrum an aircraft designed specifically for offensive or defensive combat operations possesses characteristics which set it apart from those seen regularly arriving at and departing from the world's airports, but aeroplanes and helicopters designed for other military activities, such as reconnaissance, transport or in-flight refuelling, may resemble their civilian counterparts quite closely.

INFLUENCES ON MILITARY AIRCRAFT DESIGN

While technology will be the focus of this study, it should be noted at the outset that technology is only one influence, albeit extremely important, on the construction and entry into squadron service of any military aircraft. National priorities in the allocation of resources may dictate a finite limit to the technological investment military aviation may receive. In the United Kingdom, for example, the technological expertise existed in the 1960s to develop a very advanced offensive combat aircraft, the TSR-2, but resource considerations played a large part in the abandonment of the project. Or a government may perceive a shift in emphasis in the nature of an external threat and modify its aircraft procurement plans accordingly. Thus, although Britain produced the first jet fighter, sixteen years later in 1957 it was decided that the manned interceptor was no longer required in the light of Russian nuclear missile development and again technological expertise was diverted elsewhere.

Conversely, a country's perception of its foreign and defensive interests may change, and technology will be focused accordingly. Thus in the years immediately after World War II the Soviet Union concentrated heavily on investment in air defence fighters, only later broadening the base of her air power by designing extended-range tactical and strategic bombers, long-range transports and vertical take-off/landing (VTOL) aeroplanes to operate from aircraft carriers. Occasionally, development of a particular kind of aircraft will be encouraged by one individual service in the pursuit of its primary mission to the exclusion of others which it regards as secondary, or the pressure of another branch of the armed forces. The development of the B-52 in the USAF's Strategic Air Command in the 1950s and the contemporary comparative lack of technological investment in tactical aircraft had such an element. But since the Wright brothers' achievement at Kittyhawk, military aviation has been at the forefront of the technological revolutions of the twentieth century: in the development of the combustion, jet and rocket engines; in the evolution of new metal structures and engineering practices; in the invention of radar; and finally, as a major stimulus to the electronic and computer explosion in the second half of this century. It is, however, significant that in the next decade the computer may drive the shape of air power technology rather than the other way round.

COMBAT AEROPLANES

The components of a military aeroplane have remained basically unchanged since the hesitant combat operations of World War I. They comprise the airframe itself, the engine or engines which power it, the instruments which allow the aircrew to operate it, and the aircrew themselves. The spirit and general attitude of today's aircrew would be recognised by their ancestors, but little else would be. The first aircraft to be handed over to the Air Battalion of the Royal Engineers of the British Army, the cradle of the Royal Flying Corps and progenitor of the Royal Air Force, pressed the frontiers of aviation technology in 1912 and was given a certificate of airworthiness to confirm it.

BE1 Certificate

> This is to certify that the aeroplane BE1 has been thoroughly tested by me, and the mean speed over a three-quarter mile course with a live load of three hundred and fifty pounds and sufficient petrol for one hour's flight is 58–59 miles per hour. The rate of rising loaded as above has been tested up to six hundred feet, and found to be at the rate of one hundred and fifty-five feet per minute. The machine has been inverted and suspended from the centre and the wings loaded to three times the normal loading. On examination after this test, the aeroplane showed no signs of defect.

> 14 March 1912   S. HECKSTALL-SMITH
>   for Superintendent Army Aircraft Factory

How different is the modern combat aircraft; for example, Panavia Tornado GR1′ whose overall layout is seen below (Fig. 1.1).

Characteristics of the Military Airframe

The airframe is not only the most obvious component of a modern combat aeroplane; it will generally cost more than half and account for almost half the empty weight of the total aeroplane. The other major cost and weight elements,

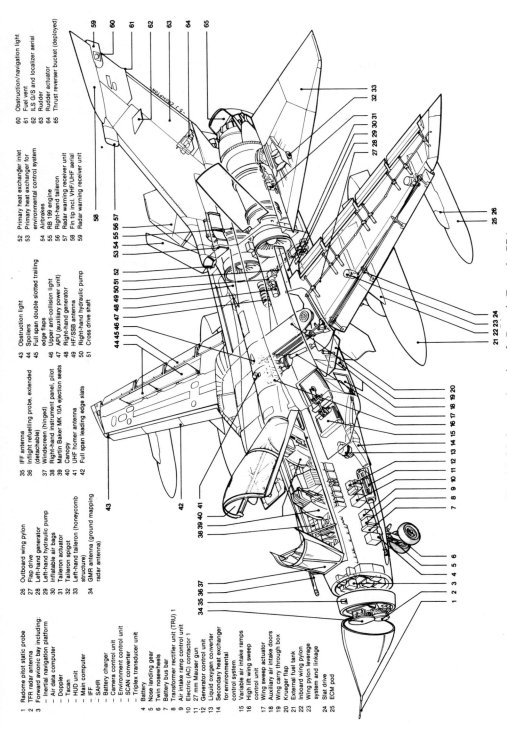

1 Radome pilot static probe
2 TFR radar antenna
3 Forward avionic bay including:
– Air data computer
– Inertial navigation platform
– Doppler
– Tacan
– HUD unit
– Main computer
– IFF
– SAHR
– Battery charger
– Camera control unit
– Environment control unit
– SCAN converter
– Triplex transducer unit
4 Battery
5 Nose landing gear
6 Twin nosewheels
7 Battery bus bar
8 Transformer rectifier unit (TRU) 1
9 Air intake ramp control unit
10 Electric (AC) contactor 1
11 27 mm Mauser gun
12 Generator control unit
13 Liquid oxygen converter
14 Secondary heat exchanger for environmental control system
15 Variable air intake ramps
16 High lift wing sweep control unit
17 Wing sweep actuator
18 Auxiliary air intake doors
19 Wing carry through box
20 Krueger flap
21 External fuel tank
22 Inboard wing pylon
23 Wing pylon leverage system and linkage
24 Slat drive
25 ECM pod

26 Outboard wing pylon
27 Flap drive
28 Left-hand generator
29 Left-hand hydraulic pump
30 Inflatable air bags
31 Taileron actuator
32 Taileron spigot
33 Left-hand taileron (honeycomb structure)
34 GMR antenna (ground mapping radar antenna)
35 IFF antenna
36 Inflight refuelling probe, extended (detachable)
37 Windscreen (hinged)
38 Right-hand instrument panel, pilot
39 Martin Baker MK 10A ejection seats
40 Canopy
41 UHF homer antenna
42 Full span leading edge slats

43 Obstruction light
44 Spoilers
45 Full span double slotted trailing edge flaps
46 Upper anti-collision light
47 APU (auxiliary power unit)
48 Right-hand generator
49 HF/SSB antenna
50 Right-hand hydraulic pump
51 Cross drive shaft
52 Primary heat exchanger inlet
53 Primary heat exchanger for environmental control system
54 Airbrakes
55 RB 199 engine
56 Right-hand taileron
57 Radar warning receiver unit
58 Fin tip incl. VHF/UHF aerial
59 Radar warning receiver unit
60 Obstruction/navigation light
61 Fuel vent
62 ILS G/S and localizer aerial
63 Rudder
64 Rudder actuator
65 Thrust reverser bucket (deployed)

FIG. 1.1. Modern combat aircraft: Panavia Tornado GR1.

excluding weapon systems, comprise propulsion units and avionics. The requirements for a combat aircraft will vary in their proportions, according to the roles for
which it is designed, but the most important are take-off and landing performance,
speed, acceleration, rate of climb, manoeuvrability, endurance, and survivability.
The ideal combat aircraft would reach the ideal in each characteristic, but in
practice the aircraft designer must effect a compromise because some can only be
achieved at the expense of others. Speed, for example, is primarily dependent on
the ratio between the thrust developed by the engines and the weight of the
aircraft, but is affected by the resistance of the operating medium, referred to as
aerodynamic drag, which is the product of the aerodynamic shape of the aircraft
itself. Endurance can be extended by increasing fuel-carrying capacity, but only to
a point beyond which weight and volume will threaten to reduce the power-to-
weight ratio and ideal aerodynamic characteristics. For this reason, among others,
in-flight refuelling has come to play an increasing role in the considerations of
designers and operational planning staffs.

The function of any combat aircraft, in offence or defence, is to carry and
deliver weapons, and consequently the contribution of technology in achieving
reductions in aircraft weight is extremely important.

Airframe Weight

Since World War II, combat aircraft have grown progressively heavier, although
airframe weight proportions have remained generally constant. Modern technology, however, now offers the promise of significant reductions. Already the use of
metal powders instead of traditional melting techniques is making possible the
production of new alloys. Titanium, for example, can now be reconstituted from
powder produced by atomising molten metal, achieving a fine microstructure
which can facilitate conventional machining and shaping. A new aluminium–lith-
ium alloy, produced by a similar technique, is 10 per cent lighter, 15 per cent stiffer
and gives greater resistance to fatigue cracking than previous generation metals.
Further weight savings, in some cases as much as 40 per cent of airframe weight,
could be achieved by the use of composite materials such as fibreglass, carbon,
boron and kevlar. They are both lighter and stronger and can be formed with
complex aerodynamic shapes impossible with traditional materials. In that
characteristic lies a further advantage to aircraft manufacturers.

Aerodynamic Construction

The new materials will not only reduce weight, but give the designer an
opportunity to construct an airframe to meet aerodynamic requirements which
have hitherto presented an unsurmountable challenge. For example, the theoretical advantages of a forward-swept wing have been known for many years. It
reduces drag, increases lift, reduces wing size and associated weight, reduces fuel
requirement and engine size and by allowing a smaller aircraft to be designed for
the same task reduces visibility and reduces vulnerability. The Luftwaffe
experimented with a forward-swept, four-jet bomber in 1944, but it was not
successful. Hitherto, the forward-swept wing has been structurally vulnerable,

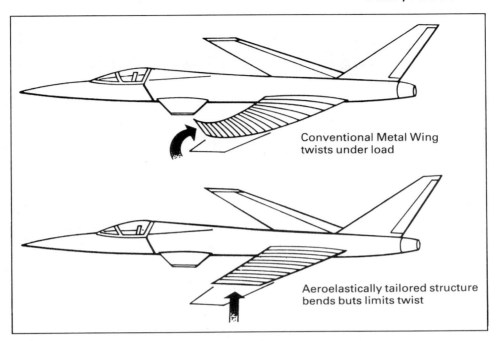

Fig. 1.2. X-29 upper and lower wing covers are formed in one piece using aeroelastically tailored composites.

because the aerodynamic flow over it has tended to twist the forward edge upwards, the trailing edge downwards, increasing lift but threatening to break off the wings from the fuselage. To construct a wing strong enough to withstand such pressure induced unacceptable weight penalties.

In the early 1970s, however, the USAF Flight Dynamics Laboratory at Wright Patterson Air Force Base began to experiment with composite materials. In 1974 Colonel Norris J. Krone Jr, USAF, proposed the construction of a forward-swept wing of composite graphite–epoxy laminates which could be tailored to twist the leading edge down while bending upward with increased loading. His belief that such a wing would prove structurally stable and reduce aircraft weight was put to the test by a Grumman designed model in 1977 at the NASA Langley Research Center (see Figs. 1.2 and 1.3). By 1984, the X-29 was flying (see Plate 1.1). Its operational effectiveness and military acceptability remained to be proven, but the promise of technology to enhance performance and reduce weight was there for all to see. A new and significant expression had entered the vocabulary of combat airframe design: 'aeroelastic tailoring'.

COMPUTER ASSISTED DESIGN

The application of the computer to aircraft design has already reduced the basic stages of drawing from months to minutes: the USAF's F-15 was one aircraft

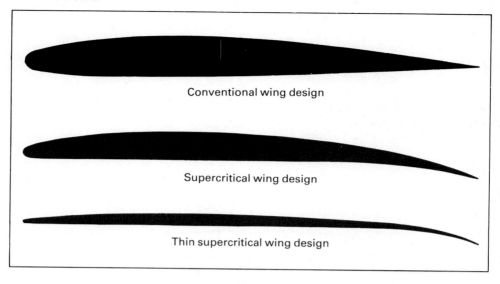

Conventional wing design

Supercritical wing design

Thin supercritical wing design

FIG. 1.3. Typical wing cross-sections—supercritical wing delays and softens shock waves on upper surface at transsonic speeds.

PLATE 1.1. Grumman X-29 forward swept wing aircraft.

which benefited from such early techniques. Even that contribution, by shortening preproduction time, reduced both costs and the period from concept to squadron service: both important considerations in the production of an advanced combat aircraft in a period of rapidly evolving technology. Now, however, of even greater significance in the design stage is the emergent technology of numerical aerodynamic simulation, or computational fluid dynamics. Digital simulation of airflow round an airframe has already greatly improved design performance. NASA's new computer systems can produce 1,000 million operations a second with 240 million words of internal high-speed memory. Although the ability to forecast the full true dimensional flow of turbulence round an airframe is still in the future, it is already possible to investigate a far wider variety of shapes before the costs of either model or demonstration aircraft are actually built. The impact of this design technology can be seen in the variety of shapes chosen by British Aerospace and Dassault in Europe, or Boeing and McDonnell Douglas in the United States, for the air superiority fighter of the future. Already, valuable information has been gleaned about the behaviour of airflow around traditionally difficult sections of the aircraft such as engine nascelles and external stores such as drop tanks, weapons or reconnaissance pods.

STEALTH TECHNOLOGY

In their quest for aerodynamic perfection, however, aircraft designers cannot lose sight of the fact that they are not designing an aerobatic display aircraft, but one which must survive in combat. As is explained in subsequent chapters, weapon technology has approached the point where an aircraft which is detected by an enemy is an aircraft in imminent danger of destruction. There are several sources and kinds of threat, and several corresponding defensive ripostes, but increasingly the airframe itself must be a factor in the reduction of its own vulnerability. In military history, an enemy has been detected by either sight or sound. Counter-measures were primarily camouflage and stealthy movement. In modern air warfare, 'stealth' is the term associated with a variety of technologies which seek to reduce detection of an aircraft not now primarily by visual and audible sensors, but along the entire electromagnetic spectrum, and especially in the microwave frequencies. Aircraft shape, size and materials used in structure are all areas in which the search goes on to reduce 'radar cross-section', a phrase which is freely used to denote not just reduction in size, but in electromagnetic reflectivity generally. Stealth technologies seek to reduce an aircraft's 'signature', by making it more difficult to be detected by radar, more difficult to be located by the heat which it radiates and, now of lesser significance, more difficult to be seen. Because of the relatively short attenuated radius of audible sound, the noise generated by a modern military aircraft is now only of concern to civilians below the flight paths in peacetime. The Rockwell B-1B strategic bomber described fully in Chapter 7 incorporates many Stealth features and is reported to present a radar cross-section 10 per cent of the B-1A and 1 per cent of the B-52. Three companies, Boeing, Vought and Northrop, have been commissioned by the US Department of Defense to build up an advanced technology bomber, to fly in 1987. In view of previous Northrop experience with aircraft in 1946 and 1947, speculation in the press suggested that its design would be a flying wing, which would not only offer minimum radar reflection but would be aerodynamically extremely efficient. Hitherto, such an aircraft would have been aerodynamically unstable. Now, in the 1980s, the interaction of electronics and airframe make the concept operationally feasible.

ACTIVE CONTROL TECHNOLOGY

It was not possible to verify the aerodynamic properties of the Northrop experimental plane because of its associated, highly classified Stealth technology.

There were, however, well-publicised experiments taking place in both North America and Britain designed to incorporate instability in an airframe to enhance its aerodynamic performance and then to use a different branch of technology to control it. Traditionally, stable flight has been achieved in an aircraft by aerodynamic design and mechanical controls operated by the pilot who either anticipates external influences or reacts quickly to them. Stability has been essential to flight safety. The aerodynamic and structural penalty of stability has been the need for a wing large enough to balance the negative lift induced by the tail, which in turn induced extra drag, larger and heavier engines and increased fuel load. If, however, the aircraft's centre of gravity could, in simple terms, be moved back behind the centre of aerodynamic lift, aerodynamic efficiency could be improved, the tailplane and wing could be reduced in size and an opportunity would be offered to install smaller, lighter engines with a reduced fuel load. Simultaneously, the aircraft would become much more agile and manoeuvrable: one estimate is an improved turning rate of 20 per cent, thereby considerably enhancing fighter combat capability.

Modern computer technology has made artificial stability possible, as Fig. 1.4 a and b shows. The flying control surfaces of an aircraft can be activated automatically by computerised aircraft motion sensors to stabilise and monitor performance many times a second. This 'active control technology' has been used experimentally on the US Space Shuttle and, among other aircraft, on the British Aerospace Jaguar GR Mk 1 demonstrator. The modified Jaguar began trials in 1981 and progressively reached the stage in 1983 when combat manoeuvring could take place with the computerised controls safeguarding the airframe limits, thereby protecting the pilot from inadvertent stall, spin or airframe over-stress. Active control technology is still in its infancy, but it offers the clearest indication of the ways in which the computer can revolutionise traditional airframe design. Electronic controls will be an integral part of future aircraft, and especially fighter aircraft, performing the functions at present discharged by hydraulic, pneumatic and mechanical systems.

ENGINES

The petrol-driven internal combustion engine was first installed in an airframe to drive a 'propeller' which accelerated air rearwards, thereby, by the principle of equal and opposite reaction, creating a propulsive force or 'thrust' which moved the aircraft forwards. In the gas turbine engine, on the other hand, air has energy added to it by heating as it passes through the engine and is then accelerated through a propelling nozzle, producing a high-speed exhaust jet. The reaction to the momentum of the gases acts on the aircraft as thrust. In the basic jet engine (see Fig. 1.5) a mechanical compressor induces an airflow during static ground running and in flight compresses it to an efficient combustion level. After compression, the air is mixed with fuel and burnt in the combustion chamber. Behind the combustion chamber the turbine extracts energy from the hot gases to drive the compressor and engine accessories. Behind the turbine the gases are ducted by a jet pipe to the propelling nozzle, being converted into kinetic energy which creates the momentum to produce the thrust.

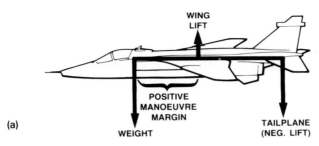

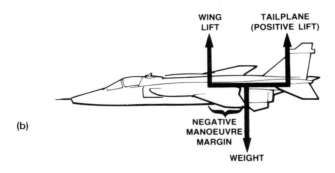

FIG. 1.4. (a) Conventional aircraft naturally stable. (b) Unstable aircraft artificially stabilised.

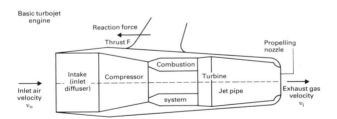

As aircraft speeds continued to increase during World War II, limitations on the piston engine/propeller power plant became apparent. Above about 450 knots turbulence created around the propeller increases drag, while the piston engine itself is proportionately heavy. Consequently, invention and development of the gas turbine, or 'jet', engine was spurred by the constant combat requirements of higher speed, greater acceleration and increased endurance. The gas turbine engine has been progressively enhanced and several modern technologies offer further, improved, thrust-to-weight ratios. Its effectiveness is illustrated in Fig. 1.6. The propeller survives in aircraft driven by a turboprop power plant, as for example in the Russian TU-95 Bear series, because the combination of gas turbine

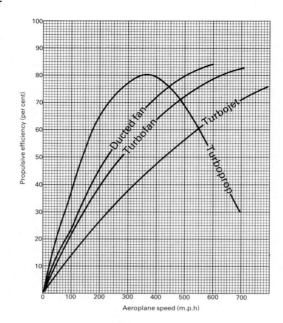

FIG. 1.6. Gas turbine engine effectiveness.

and propeller driven from it continues to give very efficient and economical thrust at subsonic speeds.

The most widely used type of engine in modern combat aircraft, the turbofan, draws on the advantages of both turbojet and propeller. Basically, in a turbofan engine the propeller has been withdrawn inside the engine intake, whereby ducting of air over the propeller tips removes the aerodynamic drag but preserves the economy and lighter weight of the gas turbine. Further efficiency is achieved by channelling a proportion of air past the combustion chamber, through one or more compressors, then feeding it into the exhaust gases downstream of the turbines so that they are cooled down; the exit velocity is proportionately reduced and the overall propulsive efficiency of the system is improved. If a further combustion stage is used, to create 'reheat', the presence of a higher proportion of oxygen in the 'by-passed' air makes the surge power generated by the reheat even more effective. Because the 'by-pass' engine is not required to deal with all the airflow from the inlet, the high-pressure compressor, turbine diameter and combustion chamber can all be reduced in size, decreasing overall engine weight and increasing the thrust-to-weight ratio as well as reducing fuel consumption. The result is a power plant which enables combat aircraft to take advantage of the turbofan's low specific fuel consumption at subsonic speeds without incurring an excessive drag penalty in supersonic flight; combining subsonic cruise with a high supersonic dash capability.

Developments in powder metallurgy and new manufacturing techniques such as directional solidification, single-crystal metallurgy, rapid solidification rates and isostatic pressing are making possible the construction of stronger turbine and compressor discs with higher temperature capacity and stress resistance. Elec-

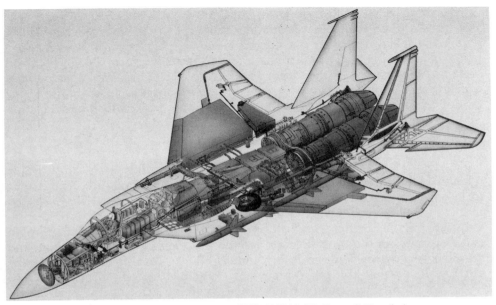

PLATE 1.2. Cutaway diagram of USAF F-15 (McDonnell Douglas).

tronic digital engine control will further improve fuel efficiency, control reliability and flight safety by automatic restriction within engine surge boundaries. As in airframe construction, computerised fluid dynamics will indicate from the design stage the nature of the airflow pattern over the sets of rotating blades and lead to further compound effectiveness throughout the engine cycle. The Pratt and Whitney F100-PW-220 engine being installed in the USAF F-15, for example (see Plate 1.2), is fully digitally controlled, while the Turbo Union Mk 104 engine to be fitted to the RAF Tornado Air Defence Variant for the RAF and to the British experimental fighter demonstrator has a full authority digital control without any back-up hydromechanical fuel control. A further development by Rolls Royce, the XG-40, seeks to achieve a 10:1 thrust-to-weight ratio and draws upon contemporary technology with a high-strength compressor and turbine discs, low-density single-crystal turbine blades, advanced sealing techniques and ceramic coatings in the combustion and turbine area.

The USAF has placed a series of contracts with manufacturers in preparation for the purchase of engines for its projected advanced tactical fighter which illustrate the spread of technologies involved in modern combat aircraft engine design. They include a one-piece hollow titanium turbofan; a nickel-based superalloy compressor; titanium alumide fuel injector nozzles and combustion cases; all composite inlet guide vanes; a graphite polyamide for the front frame of the turbofan; a two-dimensional vectoring and reversing exhaust nozzle partly constructed from carbon–carbon which will provide a heat shield and reduce the amount of active cooling required at the jet exhaust; and monocrystal blade castings with greater resistance to oxidisation. It has been claimed that the combined impact of the new processes, if successful, will generate 30–35,000 lb of thrust, compared to the Pratt and Whitney F100-PW-220's 23,830; 75 per cent of that thrust will be available without after-burner, and the engine weight will be reduced by at least 10 per cent. Achievement of such characteristics would make a major contribution to the construction of a fighter aircraft capable of high and sustained turn rates with the excess power to accelerate or climb rapidly after a combat manoeuvre. In Chapters 3 and 4 the importance of such capabilities in the face of modern air-to-air and ground-to-air weapons is explained in detail.

INSTRUMENTS AND CONTROLS

The crew's traditional need for flight information has been considerably increased with the rapid advances in combat aircraft performance. At the same time, other systems have proliferated and become more complex. The pilot, and in a two-seater aircraft his navigator or weapon systems operator, now need to know not only their height, heading, speed, angle of attack (inclination), position and fuel state, but they must operate communications, navigation, weapons and self-defence systems while probably enduring high g loading and maintaining unimpeded combat tactical awareness at very high speeds, probably at low level. Not surprisingly, manual and visual communications between man and machine, as represented by traditional instrument displays, are becoming saturated.

In aircraft such as the Panavia Tornado GR1 Mk 1 the workload of the two-man crew is reduced by partial automation of flight controls, for example in the autopilot system capable of flying the aircraft at low level, at high speed, in all weathers; and by selective automatic presentation of flight data on the head-up display (HUD) in front of the pilot. The Tornado cockpit layouts can be seen in Figs. 1.7 and 1.8. In the following generation of combat aircraft, typified in Europe by the British Experimental Aircraft Programme fighter (EAP) and by the French Rafale project, and by the Advanced Fighter Project in the United States, automation and selective displays have been more extensively incorporated. Instead of individual instruments presenting constant data, cathode ray tube (CRT) multi-functional displays are used to present data either on demand or, in the case of system malfunction or emergency, automatically. In EAP, for example, the warnings display screen remains blank until a malfunction in a system occurs, whereupon warning lights flash alongside the HUD, the CRT identifies the failed system and at the same time an increasingly stentorian lady's voice warns the pilot through his headset. Now, many vocal communications between aircrew and machine remain one-way. By 1986 an experimental voice control system had been installed in a USAF F-16. Before flight, the pilot's voice was recorded on an audio tape cassette. Thereafter the system would match the pilot's command with the prerecorded words and perform the desired function. The 'interactive voice system' may respond by presenting information by synthesised voice reply or by visual display, or initiate actions such as changing radio frequencies, navigation way points, target coordinates and radar range/modes. Thereby, the pilot will have fewer distractions from head-up control of the aircraft and less breaks in concentration on the combat environment. Flights tests have sought to evaluate voice recognition in a realistic cockpit environment where noise from the

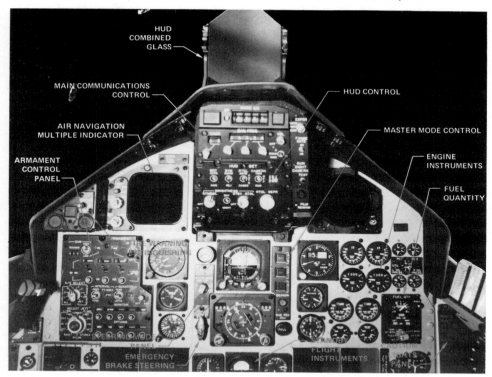

PLATE 1.3. The contemporary fighter aircraft cockpit—USAF F-15 (McDonnell Douglas).

airstream, engines, breathing, could be superimposed on the voice signal, as well as the effects of vocal distortion at various levels of g loading.

In military transport aircraft similar principles in instrument and display design have been applied, but for different reasons. The transport crew is unlikely to meet high g loading or high-speed combat demands, but in every branch of modern air operations flight crew are expensive and increasingly scarce; in the West at least. Consequently, reduction in aircrew workload can lead to reductions in aircrew requirement; as explained in Chapter 7, this situation is imminent with the development of the C-17.

As is explained in detail in Chapter 3, the demands on the air superiority fighter are such that increasingly flight controls and weapon systems are being designed with integral and coordinated automatic parameters, leaving the pilot more time to contribute operational judgement and response to his machine. At the heart of the automated displays and controls are the very-high-speed integrated circuits of airborne computers which are leading aircraft designers to move from the current stage of automatic 'fly-by-wire' flight controls (an example of which is shown in Fig. 1.9) to the direct integration of electronic signal with airframe and engine response. Indeed, the major hurdle may lie not in the technology itself, but in the software designed to make the computer work. The more complex the system has to be, the more circumstances it has to identify and respond to, and the easier it is for a fault to remain undetected for a considerable period. As a result, extensive

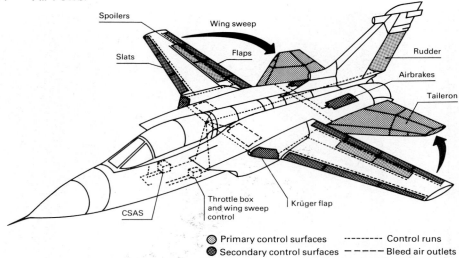

Fig. 1.9. Tornado 'fly-by-wire' control system.

programmes are under way to investigate fault occurrence and reliability in integrated avionic systems. The Avionics Integration Research Laboratory at Langley Air Force Base in the United States is able to simulate an aircraft and its environment, while a converted Lockheed Hercules C-130 'high technology test bed' is being used to evaluate both specific transport technologies and more general interactive electronic and avionic systems.

Looking further ahead, fibre optics may supplement or replace electronic transmissions as signal conductors. A signal-bearing light beam is invulnerable to electromagnetic interference, either from other systems or external sources; it can carry thousands more signals than a copper wire of comparable weight; it does not radiate tell-tale emissions and it is invulnerable to electromagnetic jamming. By 1985, a proposal had been made for a 'fly-by-light' control system for the US Army's experimental light helicopter programme.

VERTICAL LIFT AIRCRAFT

Several factors in modern warfare have combined to enhance the value of aircraft which can take off and land vertically. The ability to operate from small, unmade airfields, roads, clearings and virtually any flat space large enough to accommodate its wing span has made the British Harrier a valuable close support weapon platform. Any military aircraft which does not need to rely on long concrete runways offers many more flexible options to its commanders and reduces its vulnerability to counter-air operations against airfields. With increased reliability and power-to-weight ratios, the conventional helicopter is also assuming much greater tactical significance. Computerised design techniques have been applied equally to helicopters and fixed wing aircraft, with resulting improvements in rotor blade aerodynamics, while composite materials described earlier have been applied to rotor blades with subsequent improvement inefficiency and

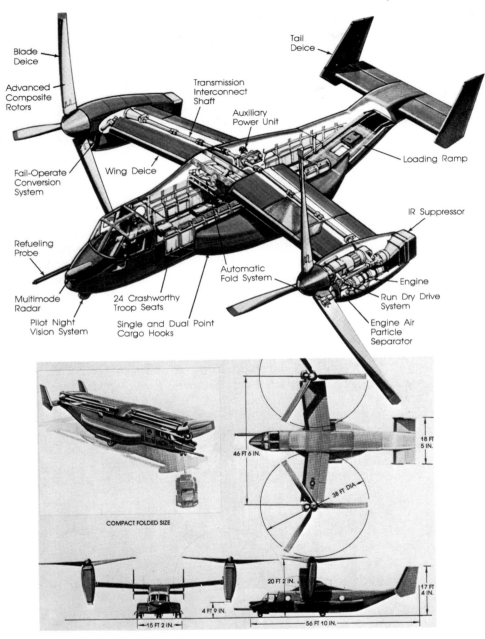

Blade
Deice

Advanced
Composite
Rotors

Tail
Deice

Transmission
Interconnect
Shaft

Auxiliary
Power Unit

Loading Ramp

Fail-Operate
Conversion
System

Wing Deice

IR Suppressor

Refueling
Probe

Automatic
Fold System

Engine

Run Dry Drive
System

Multimode
Radar

24 Crashworthy
Troop Seats

Pilot Night
Vision System

Single and Dual Point
Cargo Hooks

Engine Air
Particle
Separator

COMPACT FOLDED SIZE

46 FT 6 IN.

38 FT DIA.

18 FT
5 IN.

20 FT 2 IN.

17 FT
4 IN.

4 FT 9 IN.

15 FT 2 IN.

56 FT 10 IN.

FIG. 1.10. V-22 Osprey multimission tilt rotor.

reduction in weight. Meanwhile, improvements to engine performance in fixed
wing aircraft have been incorporated, where relevant, to the turboshaft engines
which drive the gearbox, which in turn powers the helicopter's rotor blades.
Nevertheless, conventional helicopters remain restricted cost effectively to a

speed of some 200 knots, primarily because of the aerodynamic properties of the rotor blades, either from advancing blade supersonic flow or retreating blade stall. As a result, several efforts have been made to combine a fixed wing aerodynamic shape with a propulsion unit capable of providing vertical lift.

One of the most widely publicised development programmes is the Bell/Boeing V-22 Osprey twin-engine tilt rotor aircraft, shown in Fig. 1.10, intended for use by all four US armed services. The experimental demonstrator which flew in 1977 on the Bell/NASA XV-15 was built from conventional materials, but Osprey draws extensively on composites. Twin turboshaft engines on the wing lateral axis at each wing tip give vertical lift, then swing 90 per cent forward to function as conventional turboprops. Osprey will have a maximum vertical take-off and landing (VTOL) weight of 40,000 lb and short take-off and landing (STOL) of 50,000 lb. It promises an unrefuelled ferry range of 2,100 nautical miles and a cruising speed well in excess of 300 knots. It is the product of computer aided design and manufacturing and more than 5,500 hours of wind tunnel testing of nine various scale models. Without the carbon composite materials, which comprise 60 per cent of the aircraft's weight, it would not have been possible to build a wing big enough and strong enough to withstand the heavy loads of the two engines and 38 ft diameter rotor blades set so far from the aircraft's centre line and forward of the wings' lateral axis. The product promises to be a combination of cost effective tactical transport, with the additional operational flexibility of the helicopter at ranges and speeds akin to those of the C-130 Hercules, albeit without comparable payload. The unique aircraft is designed to provide combat search and rescue for the US Navy, special clandestine operations for the US Air Force, medium assault transport for the US Marines and multi-role tactical transport for the US Army. In the space of one decade, rapidly advancing computer, microprocessor, electronic, engine, aerodynamic and composite material technologies have combined to translate a futuristic concept into a practical military aircraft with the potential to discharge many different support roles. The same technologies, in variable proportions, have powerfully influenced the evolution of aircraft, weapons and associated systems in every traditional role of air power.

2

Air Defence: (1) The Home Base: Early Warning, Interception and Surface-to-Air

THE PRINCIPLES

Modern defences against attack from the air have their roots in two quite different traditional military principles. One is inherited from land warfare: the security of the home base; the other from naval warfare: command of the sea, translated into the third dimension as 'command of the air'. In the World War I infancy of air warfare the security of the British home base, for example, was threatened by attacks from German four-engined Gotha biplane bombers. The threat was taken so seriously by the government of the day that an independent third service, the Royal Air Force, was created with a specific responsibility for 'air defence' of the British mainland, hitherto a marginal responsibility of the Royal Navy. Meanwhile, over on the European continent, the aeroplane was adding a further dimension to the war in the trenches. First used for artillery spotting and reconnaissance, it was quickly recognised by the opposition as a dangerous nuisance, and rudimentary attempts were made to get rid of it, by shooting it down by gunfire either from the ground or from other aircraft. Denying the use of the sky to the enemy, and preserving unimpeded access for one's own aircraft, became the twin objectives in the struggle for 'command of the air'. Subsequently the breadth of implication in that expression became tactically refined, as 'air supremacy' or 'air superiority', to denote 'command of the air' in a particular region for a particular period. By the second half of the twentieth century, command of the air or its local derivative became essential not only for the success of any operation in the air itself, but also for the integrity of activities on land or at sea. The two principles have gradually merged, as command of the air will obviously enhance the security of the home base as well as make other kinds of air operations feasible. But even after seventy-five years the two have distinct requirements, in addition to those in common, and the distinctions are equally evident in the provision for them in aircraft, weapons and associated systems.

17

AIR DEFENCE COMPONENTS

Early writers on air power exaggerated both the destructive capacity of the manned bomber and its invulnerability. The foresight of men like Douhet, Mitchell and Trenchard should perhaps have been tempered by a greater appreciation of the practical difficulties in designing and building aircraft large enough to carry sufficient weight of explosives, incendiaries and gas weapons. They might have foreseen problems of navigation and limitations of weather. But given the disproportionate impact of air attack on civilian and ground force morale, and without being able to anticipate the invention of radar, it really did seem that 'the bomber would always get through'. Quite simply, by the time the presence of an enemy aircraft was detected either audibly or visually, it was too late for the primitive defensive fighters to climb away from their airfields with any hope of an interception. They had neither the engine power to enable them to climb swiftly nor the instruments to help them locate an enemy beyond visual range. Combat air patrols could be flown along likely intruder routes; searchlight batteries might occasionally illuminate a bomber long enough for ground fire and any interceptor in the area to attack, but the odds until World War II remained with the bomber and only allowed spasmodic contact with the fighter.

Since the invention and deployment of radar, however, air defences have become large-scale, highly complex, closely integrated systems in which the manned aircraft has become only one component. The components are the same, whether providing for air defence of the home base, or for tactical air supremacy associated with simultaneous warfare on land or at sea, but their relative significance and detailed elements vary and, consequently, so does the associated technology. The system comprises:

(1) Early warning of air attack.
(2) Communication of attack details to active defence units, including manned aircraft (interceptors) and surface-to-air defences (SAD).
(3) Interceptors.
(4) SAD.
(5) Control of interceptors and SAD.
(6) On-board interceptor target acquisition and identification equipment.
(7) Weapons.
(8) Plus complementary but quite separate offensive counter-air operations.

(*Note*: Every stage of the interception process is now shrouded in electronic warfare, which is dealt with as a separate subject in Chapter 5 below.)

EARLY WARNING

The most famous episode in the history of air warfare concerning early warning occurred in World War II during the Battle of Britain. A chain of radar stations hastily constructed around Britain's south and east coasts during the previous couple of years allowed incoming German bomber squadrons to be located, and their direction anticipated, before they reached British airspace. RAF fighter squadrons were scrambled and directed to interception points, thereby achieving the maximum of surprise, concentration of force and, ultimately, success in the Battle of Britain.

Important as early warning of air attack was in 1940, its significance in modern warfare is difficult to over-exaggerate. Bomber and fighter bomber aircraft now possess greatly increased speed, an ability to place freefall and guided air-to-surface weapons with considerably enhanced accuracy and destructive power and above all an ability to launch their weapons at increasingly distant ranges from their designated targets. The devastating military advantages accruing from a surprise air attack were evident at Pearl Harbor and in Russia in 1941 and were repeated in the Middle East in 1967 and to a lesser extent in 1973. An initial large-scale air offensive is believed to comprise an essential ingredient in Soviet military planning for offensive operations against the Western alliance. Consequently, the provision of early warning of enemy air attack is essential both to reduce the likelihood of a surprise attack and to concentrate air defences in time and space to the greatest destructive effect.

Ground-based Systems

Traditionally, early warning of air attack has been provided by ground-based systems. After World War II, the Soviet Union was the first to deploy new chains of radar stations along her Baltic, Arctic, western and to a lesser extent eastern frontiers to guard against the perceived threat from USAF intercontinental and medium-range bombers. With the growth of Soviet long-range aviation in the 1950s, both Britain and the United States responded by building similar chains across the North American continent, up the east coast of Britain, and across Western Europe. But whereas Soviet development was steadily maintained, Western modernisation was given lower priority, as the Soviet Union did not proceed with her expected development of her long-range bomber force but, under the direction of Khrushchev, emphasised instead the production of medium and intercontinental ballistic nuclear missile forces. Since the early 1970s, however, the Soviet Union has modified her military doctrine to place greater emphasis on conventional warfighting capability and has introduced a new generation of bombers. These, including Tu-22M (Tu-26?) Backfire, Blackjack and Tu-95 Bear-H, can carry both freefall and stand-off air-to-surface weapons with either conventional or nuclear warheads. Consequently, the subsequent decade has seen a major re-emphasis on the construction and deployment of new ground and airborne early warning systems in the West.

Three separate but ultimately integrated, or 'netted' in the military jargon, ground-based systems are being developed in North America, continental Europe and the United Kingdom. In North America, the old series of early warning stations stretching over Canada to the Arctic Circle is being modernised, while four new bases, drawing upon new radar techniques, are to be built in the United States itself. In Alaska and Canada the renamed 'North Warning System' will be equipped with thirteen long-range (370–460 km) AN/FPS-117 radars built by General Electric and thirty-nine short-range (110–150 km) built by Sperry. The latter offers the additional advantage of low-level coverage and remote control, thereby reducing manpower requirements in particularly inhospitable terrain.

Whereas these are 'conventional' radars, using line-of-sight radio emissions and reflections, albeit enhanced by computerised integration, target identification and location, the four new installations in the United States mark a greater step in radar coverage. OTH-R radars work on the principle of bouncing high-frequency transmissions off the ionosphere and picking up reflected returns from targets as much as 3,000 km away, well over the visual horizon—previously the limiting factor on radar detection range. The technique demands extremely complex signal and data processing. The transmitter aerials are 1 km in length and the receiver antennae 1.3–1.5 km, while 160 km physical separation is required between transmitter and receiver arrays. As the minimum detection distance is 900 km, it is apparent that the system is for intercontinental rather than tactical application in a theatre of operations. Nevertheless, the overlapping coverage provided by the two new conventional radar systems and the four OTH-R units should reduce considerably the threat of a surprise air attack on North America and enhance by a similar proportion its air defence.

In Europe, however, while the fundamental need for early warning is equally important, the geostrategic circumstances differ considerably. In any military confrontation between East and West, there would be no vast expanse of Arctic waste for a pre-emptive air attack to traverse before reaching its targets. The new generations of heavy Soviet bombers have the range to drop freefall weapons on almost all mainland Western Europe flying at low level from bases in Eastern Europe or Western Russia. If they should carry air-launched missiles such as the AS.15TT or later generation cruise missiles, the incursion they would need to make into Western airspace would be even further reduced. Moreover, the threat of medium and heavy bomber attack on home bases in Western Europe and Britain is not the only kind of air attack for which early warning is required. Central Europe, the northern and southern flanks of the NATO alliance would probably become theatres of conventional conflict on land, to which air power would make a major contribution. The fighter bombers of Soviet Frontal Aviation, together with helicopters under army command and light bombers from Soviet Air Force regiments, would be seeking to discharge their tactical roles of 'accompaniment': air support of a deep and wide-ranging armoured offensive. Early warning of the location, direction and size of such aerial accompaniment would be essential to the successful provision of air defence by the Western Alliance. The steady but considerable increase in Soviet tactical air offensive strength has made the battle for air supremacy in Western Europe a critical element in any major confrontation, a fact recognised by both sides. Under such circumstances, early warning of attack on the home territories of the Western European allies overlaps its contribution to the achievement of tactical air superiority in a combat theatre. As a result, a major modernisation programme has also begun in the NATO Air Defence Ground Environment System (NADGE). In Germany four new air defence radars are being built, primarily by the Hughes Aircraft Company, and three more in Norway. The new NADGE system is designed from the outset to receive data from radars with overlapping adjacent coverage. Fusion of the surveillance plots will not only enhance the continuity and accuracy of target location and tracking, but allow control to be maintained if any one source should be lost. Traditional two-dimensional surveillance radars are being replaced by high-performance three-dimensional radars providing plot/track and height information in digital format. Thorn-EMI is developing integrated information systems for dissimilar radars unattainable before the advent of microprocessors and powerful minicomputers (see Fig. 2.1). Now one controller can use the same multifunction display for several purposes, including detection, surveillance, identification, tracking or weapons control. Meanwhile, forward of the Atlantic theatre and in the rear of the Central Region, are the four million square miles of air space which form the responsibility of the Improved UK Air Defence Ground Environment (IUKADGE) (see Plate 2.1). Advanced computer software and equipment will provide controllers with real-time integrated information from a variety of sensors, the whole compatible with the North Warning System and NADGE. As a result, ground radar stations across the whole of the NATO area will be able to provide faster, more accurate, more comprehensive and more reliable early warning of hostile air encroachment. But the same fundamental laws of physics will continue to apply: conventional radar wave propagation is restricted to line-of-sight and hence impeded by hills, cities and any other physical obstruction between transmitter/receiver and the theoretical horizon. Consequently, use is made wherever possible of high ground for locations, or raising the level of the radar itself as, for example, in the case of one of the new German installations placed in an eight-storey high radome. Modern technology, however, now offers a further alternative: mounting the radar in airborne platforms. Such developments, also in conjunction with the microprocessor and minicomputer, are perhaps the most important, if not the most obviously spectacular, in the impact of air power on modern warfare.

Airborne Early Warning (AEW)

The use of the air for reconnaissance predates the current sophisticated AEW aircraft by two centuries. The French revolutionary armies used balloons as observation posts for their artillery in 1794; in the following century the Union armies used them in the American Civil War. In both world wars, photographic reconnaissance was widely used and by the early 1960s the Super-Constellation civilian airliner airframe had been fitted with a large radome to add radar surveillance to the previous optical techniques. This aircraft, the EC-121, was originally procured to contribute to the air defence of the North American continent, but in the war in South-East Asia a later model was used to provide limited early warning of fighter interceptions against B-52 raids on North Vietnam. It could not detect, locate and track targets obscured by surface returns (clutter) and its limitations stimulated the development and production of what is now the largest and perhaps the best known of the contemporary AEW aircraft, the Boeing E-3A, known as the 'airborne warning and control system' or, more popularly, 'AWACS' (see Fig. 2.2 and Plate 2.2).

Roles

AWACS has three primary roles. First, in strategic air defence, to supplement North American air defences by providing surveillance, command and control facilities for detection, identification, tracking and interception of airborne

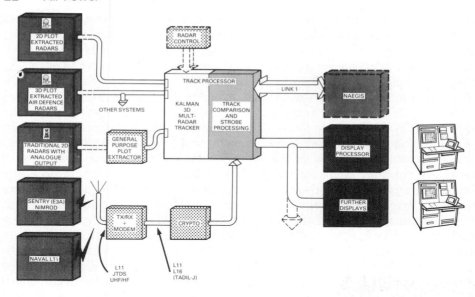

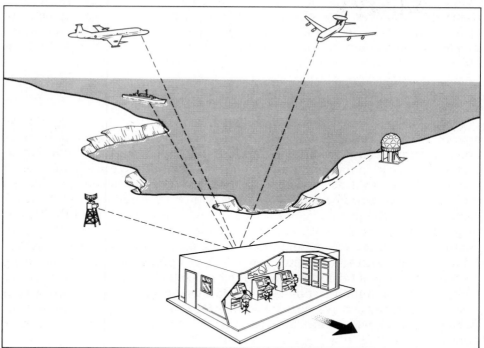

FIG. 2.1. The integration of information from different early warning systems.

attacks. Second, in a tactical European environment, it provides quick reaction surveillance, command and control communications necessary to manage effectively offensive and defensive tactical aircraft. In addition, it can detect, identify and track low-flying hostile aircraft over all kinds of terrain and control friendly

PLATE 2.1. Two RAF officers with some of the displays which will be installed in the command and control centres of the improved UK Air Defence Ground Environment (Marconi).

aircraft in the same area. Thirdly, its strategic, self-contained mobility facilitates its worldwide deployment to sustain crisis management. Its successful discharge of all three roles is the product of recent developments in radar technology and especially, yet again, of the microprocessor and computer revolution.

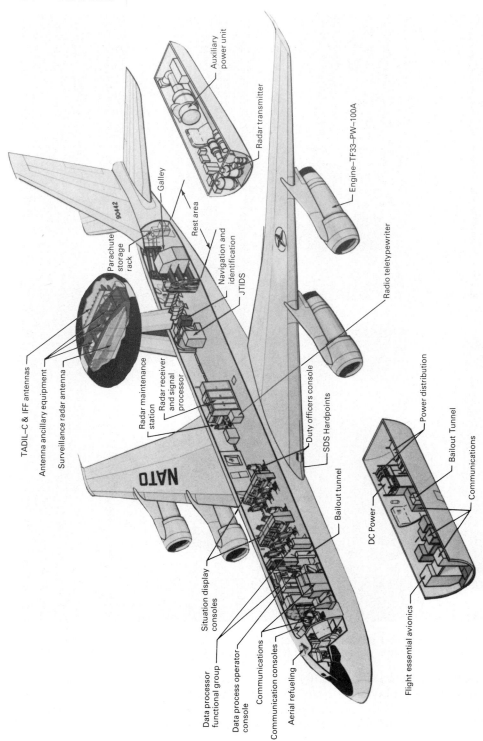

Fig. 2.2. Boeing E-3A (standard).

TADIL–C & IFF antennas

Antenna ancillary equipment

Surveillance radar antenna

Parachute storage rack

Galley

Rest area

Navigation and identification

JTIDS

Radar maintenance station

Radar receiver and signal processor

Duty officers console

SDS Hardpoints

Bailout tunnel

Auxiliary power unit

Radar transmitter

Engine—TF33–PW–100A

Radio teletypewriter

90442

NATO

Data processor functional group

Data process operator console

Situation display consoles

Communications

Communication consoles

Aerial refueling

Bailout tunnel

Flight essential avionics

DC Power

Power distribution

Bailout Tunnel

Communications

PLATE 2.2. A NATO Airborne Early Warning E-3A.

Radar Technology

The basic radar innovation was the application of 'pulse Doppler' techniques to the earlier continuous wave propagation which had measured the time taken for a radio wave to reach a 'target', be reflected, and received, thereby indicating a bearing in azimuth and a distance. By applying Doppler techniques to continuous wave radar, it had been possible to determine the radial velocity of a target while minimising the effects of ground clutter. By dividing the continuous wave into a stream of separate radar emissions or pulses, it also became possible to measure a target's range or, at higher pulse repetition frequencies, a number of possible target ranges. Finally, by using a number of different pulse repetition frequencies, ambiguous target ranges could be resolved and the true range determined. The electronic principles and detailed equipment in that evolution will be examined in a later volume in this series; of current significance is its dependence on computer technology. The microprocessor reduced the system's weight, volume and power requirements and provided for the very-high-speed integration of many data inputs and outflows, many more than simply those related to radar signal emission and collection.

The computer aided design, manufacture and test of all elements of the system. For example, the antenna array face preceded from computer design to programmed machining without the intervention of drawings or operator. The radar system performance was determined through a computer simulation of performance parameters to be expected in the air. The radar itself is tested automatically, automatic switching to standby units occurs when components fail and faults can be isolated in the air or on the front line. But the scale of such redundancy and self-testing hardware has prompted a software requirement of enormous proportions:

16,000 instructions to control radar operations; 35,000 instructions for the built-in test which continuously monitors the performance of the radar; and 125,000 instructions for fault isolation tests.

Capability

In total there are some 600,000 components in the AWACS radar. Together they give an impressive and comprehensive capability. From its normal operating height of 30,000 ft, AWACS will have a theoretical radar horizon of some 400 km, within which it can detect low-flying aircraft. Surveillance and detection ranges above the radar horizon are of course considerably greater. Open press reports have asserted that AWACS can register 600 aircraft tracks with surveillance volume, divided if necessary into thirty-two azimuth subsectors. Each subsector can have its own operating mode: pulse Doppler radar for aircraft surveillance over land; pulse radar for extended-range surveillance where ground clutter is in the horizon shadow; maritime, when a very short pulse is used to decrease sea clutter and a digital processor blanks land returns by means of stored maps of land areas; and passive, when transmitters in any subsector can be closed down, while the receivers continue to receive and process data. Electronic control of the radar beam ensures stability during aircraft manoeuvres and confines both the vertical and lateral extent, thereby permitting height measurement, accurate target location, high resolution between closely spaced targets and reducing vulnerability to electronic counter-measures. The signal processor computes the azimuth, elevation, range and range rate of each return, as well as correlating returns on subsequent sweeps and eliminating ground clutter. All data is then passed to the data processor for reference to the aircraft's own position and velocity before being prepared for display at the control consoles (see Plate 2.3). A further antenna mounted in the AWACS rotodome with the main radar supports the Identification: Friend or Foe (IFF) subsystem which provides azimuth and range information on targets with transponders. The IFF subsystem data is also fed into the tracking computer and hence to the main system displays.

Other AEW Systems

Not surprisingly, AWACS is an expensive system. In 1985 its cost was estimated at $150 m, plus the support costs of a twenty-two-man crew on board the NATO aircraft. The four Pratt and Whitney TF-33 turbofan engines give the converted Boeing Model 707 airframe 84,000 lb of thrust, an endurance in excess of 11 hours unrefuelled and a speed of 500 mph. With similar capabilities the British Nimrod AEW Mk 3 will complement E-3A as part of NATO's integrated AEW force. Nimrod is equipped with two radomes in nose and tail to provide 360° coverage and, like E-3A, has a primary pulse Doppler radar, digital data handling and automatic tracking systems integrating information from radar, IFF and other sources. Unlike E-3A, it is optimised for surveillance over water and is equipped with passive detection sensors.

The third comprehensive AEW specialist aircraft, produced originally for the US Navy, is the twin-engined Grumman E-2C Hawkeye (see Plate 2.4). Hawkeye

PLATE 2.3. A US communications operator (foreground) aboard an E-34 AWACS early warning aircraft communicates with an AEGIS ground command centre through a digital display as part of the Joint Tactical Information Distribution System (JTIDS). A West German communications technician monitors the JTIDS control panel (*background*), while a Canadian computer operator reviews the JTIDS text on the digital display. Although some communications between the AWACS and the ground stations is vocal, the vast majority of commands provided to the aircraft is through the digital display to minimise the potential misunderstanding of spoken orders (Hughes Aircraft Company).

is much smaller than the E-3A, carries three systems operators as opposed to thirteen or more, has an operating height lower than that of the E-3A, a lower speed and a shorter endurance. Consequently, Hawkeye cannot cover either the geographical area or the volume of airspace for which E-3A was designed. Nevertheless, its functions in a maritime environment are extensive, including tracking of enemy aircraft, ships and missiles; the control of carrier-launched strike/attack aircraft; intercept control of fighters, identification and clarification of hostile radar emissions; air traffic control for friendly aircraft and coordination of search and rescue activities. The additional value of its overland capabilities were graphically demonstrated by its contribution in 1982 to the Israeli annihilation of Syrian aircraft over the Bekaa Valley, when it monitored airspace both over the battle area and beyond into Syria and provided vectoring and combat information assistance to Israeli interceptors. Hawkeye's protagonists emphasise its short field capability, the by-product of its aircraft carrier origins, which confers a flexibility not necessarily enjoyed by the E-3A which requires full-

PLATE 2.4. USN Grumman E-2C Hawkeye Airborne Early Warning Aircraft (Grumman Aerospace Corporation).

length conventional—and increasingly vulnerable—runways. It is also claimed that UHF band radar gives better small target detection and copes effectively with relatively slow-moving targets. Moreover, in its twenty-year service with the US Navy, the aircraft's automated systems have constantly been updated to maximise the relatively small airframe space available and provide the greatest capacity to the three operators. It is claimed that Hawkeye can thus, like E-3A, manage 600 plus tracks.

Smaller Systems

But not every country either needs, or could afford to sustain, a fleet of such aircraft. Consequently, several other aircraft manufacturers are producing smaller, less costly, systems which do not necessarily provide the comprehensive capabilities of the E-3A, Nimrod or E-2C.

The British Royal Navy, in the aftermath of the Falklands War, has deployed a development of the Thorn EMI Searchwater radar on Sea King helicopters to provide target detection at ranges of over 100 miles from a normal operating height of 10,000 ft (see Plate 2.5). AEW Sea King can provide early warning and updates on multiple threats in the air and on the water. As with the more sophisticated systems, digital control, computer selected pulse repetition frequency, advanced signal processing and narrow beam width combine to facilitate the provision of early warning to the fleet considerably beyond ships' radar horizons. Subsequently, if required, AEW Sea King could control intercepting Sea Harriers by UHF radio link and relay battle management information reinforced by the AEW IFF system. Elsewhere, Lockheed is conducting trials with the well-proven C-130 Hercules airframe and the British General Electric APY-920 radar

PLATE 2.5. Sea King helicopter with Searchwater AEW radar (Thorn EMI Electronics).

as installed in AEW Nimrod to seek a cheaper AEW system attractive to the many countries which already operate the transport variants of the aircraft.

COMMUNICATIONS

Reconnaissance, whether pursued from a balloon or by the integration of radar and computers in the specialist role of airborne early warning, is only a means to an end, not an end in itself. Timely acquisition of data is the first step; relaying it to the battle commander and, in AEW, to the weapon systems which are to respond to the threat is the next and equally important. In Western Europe the NATO commander needs to know where and when to launch his defensive response: either interceptors or surface-to-air missiles. Consequently, the next link in the air defence chain is the communications net between radar, air defence commander and the fighters or missiles themselves.

In NATO the major communication system is known as the Joint Tactical Information System (JTIDS), which provides a secure and jam-resistant voice and digital link between the AEW aircraft and ground centres. Data messages are transmitted from Nimrod or E-3A over a coded UHF radio link, or, exceptionally, in clear language. The application of computer technology has resulted in a communication system which can use time and frequency spectrums to increase data capacity and increase jam resistance. Data can be received at more than forty airborne early warning/ground environment integration segment (AEGIS) ground stations in Western Europe and immediately processed through computers for automatic relay to display consoles in NATO air defence ground centres. There the AWACS-derived information is integrated with that from all other early warning sources, and the air commander is prepared to commit his forces to action.

THE FIGHTERS

Traditionally, aircraft designed to destroy their enemy counterparts in the air have been referred to as 'fighters', but with the development of national air defences, designed to protect a homeland rather than contest air superiority in a tactical theatre of operations, two quite distinct types of 'fighter' have evolved. Both will depend heavily on the product of the radar defences and umbilical communications to assist them enter and sustain air combat; they share many airframe and weapon system characteristics, but role specialisation strongly influences their design. At one end of the fighter spectrum is the 'interceptor', designed primarily to intercept long-range hostile aircraft whose own mission is more likely to be strategic or deep tactical bombardment of ground or naval targets rather than to dispute air superiority in the interceptor's own airspace. Nevertheless, such offensive attacks could be accompanied by long-range fighter escort and the interceptor must also be able to defend itself in air combat. At the other end of the scale is the fighter designed primarily for air-to-air combat, probably in a highly populated airspace over a theatre of ground operations. In other aircraft, such as the USAF F-15, there is a convergence of capabilities made possible by advanced airframe design and contemporary air-to-air weapon systems.

The Interceptor

The 'interceptor', characterised by the USN F-14 Tomcat and the RAF F Mk2 Tornado seen in Plates 2.6 and 2.7, has long range, the endurance to mount combat air patrols, carries autonomous radar and target-identification systems, and can intercept with both long- and short-range weapons. Airframe agility may well be of less priority in design than a high rate of climb and rapid acceleration, because modern weapons and sensors have increased the likelihood in the air defence environment of engagements beyond visual range. Nevertheless, interceptors must also be able to look after themselves and are armed accordingly.

PLATE 2.6. USN F-14 Tomcat carrying AIM-54 Phoenix air-to-air missiles (Hughes Aircraft Company).

PLATE 2.7. Tornado F2 interceptor carrying four Skyflash and two Sidewinder missiles (British Aerospace).

The F2 variant of the Panavia Tornado depicted in Fig. 2.3, for example, is twin-engined and carries a crew of two: pilot and navigator/weapon systems operator. From take-off to 30,000 ft takes less than 2 minutes, but it still has the endurance to patrol for 3 hours 300 miles from its base without in-flight refuelling. Flight efficiency at speeds from 135 knots to Mach 2.2, while retaining short runway performance is facilitated by variable geometry wings which have a leading edge sweep of 25° in the forward position and 67° when fully swept. The interceptor may be scrambled in response to a particular threat identified by radar and passed via the command centre, or it may be on a combat air patrol in the vicinity of a possible attack. The crew could receive further information while in the air from a ground or airborne controller, or they may be required to intercept without further assistance. Unless it is already within visual range of its target, the interceptor must then rely on its on-board fire control radar to acquire it and reach its missile-launching position.

Avionics

The Foxhunter radar installed in the Tornado F2, and the AWG-9 of the F-14 Tomcat, have detection ranges in excess of 100 miles. Both can track a number of targets while continuing to scan for others. Foxhunter, like the much larger radars deployed on AEW aircraft, has a variable-pulse repetition frequency to ensure maximum detection and acquisition range against approaching targets plus short-range automatic acquisition for air combat. F2 is linked to the other elements in the air defence system by JTIDS and all signal processing is handled by digital circuitry and associated microtechnology. The impact of the microprocessor in the restricted available space of an interceptor is exemplified by the updating of

TORNADO ADV

1 Pitot static probe
2 Mauser 27 mm Gun
3 Radar altimeter
4 TACAN
5 Upper IFF antenna
6 Approach aids interface unit
7 Interface unit 1
8 AOA sensor
9 Air-to-air refuelling probe
10 Ejection seat
11 High lift and wing sweep control unit
12 Wing sweep actuator
13 UHF homer antenna
14 Leading edge slats
15 Spoilers
16 Trailing edge flaps
17 Wing box
18 Upper anti-collision light
19 Auxiliary power unit
20 Right hydraulic pump
21 Right generator
22 Cross drive shaft
23 Primary heat exchanger
24 Airbrake
25 RB 199 – MK 104 engine
26 ILS antenna
27 Rudder
28 Upper TACAN antenna
29 Upper V/UHF antenna
30 ECM fairing
31 Navigation light
32 Fuel vent/dump outlet
33 Rudder actuator
34 Thrust reverser bucket (deployed)
35 Airbrake actuator

36 Taileron actuator
37 Taileron spigot
38 Taileron
39 Left hydraulic pump
40 Left generator
41 Flap drive
42 Formation light
43 Obstruction light
44 Slat drive
45 Underwing tank

46 Sidewinder AIM-9 L
47 Landing lamp
48 Air intake auxiliary doors
49 Navigation light
50 Variable air intake ramps
51 Air intake ramps control unit
52 Main engine control unit 1
53 Pyrometer amplifier
54 Life recorder
55 Vibration amplifier

56 NL/NH Governor
57 Missile programming unit
58 Sky Flash MRAAM
59 Low converter
60 AICS 1
61 IFF Interrogator
62 TV tab waveform generator
63 Engine health monitor
64 V/UHF transceiver
65 AC contactor 1

66 Generator control unit 1
67 IFF auto code change
68 Transformer rectifier unit 1
69 Battery
70 Battery bus bar
71 Lower TACAN antenna
72 Lower UHF antenna
73 Battery charger
74 Inertial navigator
75 Radar antenna

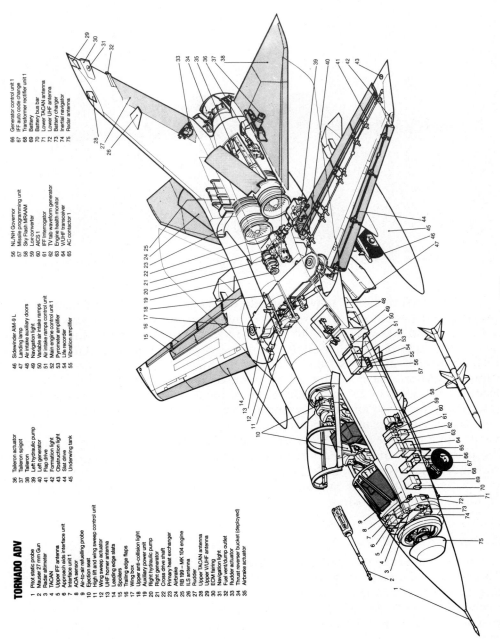

FIG. 2.3. Tornado ADV.

PLATE 2.8. An example of an airborne radar fitted in the nose of an interceptor. This is the
Thomson-CSF RDI/Electronic Serge Dassault pulse Doppler radar installed in a prototype
Mirage 2000. The radar antenna is 674 mm in diameter (Thomson-CSF).

the F14's analogue systems to contemporary digital avionics. Two airborne
computers will be able to perform two million operations per second and provide
up to 704,000 words of memory. Several hundreds of miles of wiring will be
removed from each aircraft, together with the cumbersome analogue kit. The
result will be to expand on-board data processing, improve controls and displays,
provide more effective weapon management and greatly decrease the cost of
integrating future weapons and systems. In addition, modern microprocessor
technology will improve the processing capability of the AWG-9 by a sixfold
increase in data and signal capacity (Plate 2.8).

Weapons

Finally, when the combined efforts of early warning, communications, airframe, engines, aircrew and AI radar have played their part, the end product of all
the multi-million pound systems is within grasp: the destruction of the hostile
aircraft or, in the near future, the cruise missile. The modern interceptor may be
armed with three kinds of weapon: radar-guided missiles, heat-seeking missiles, or
guns, or any combination of the three. The rules of engagement constraining an
interceptor pilot will vary according to the nature of the threat or combat

environment. Whatever they are, they will seek to ensure that he attacks only an enemy, while at the same time recognising the dilemma arising from the fact that in an age of guided weapons with a high kill probability the advantage of firing the first shot may be decisive. Hence the importance of IFF systems carried by all interceptors. They are usually variations of a radar emission transmitted so as to trigger a response from a 'transponder' in the illuminated aircraft. Or the interceptor's own radar warning system will alert the crew to the fact that they themselves are being illuminated by a surveillance, air intercept or even homing weapon radar, suggesting that the emitting source is less than friendly. If, on the other hand, AEW or other sources have established that aircraft in a particular sector are all hostile, then the interceptor can achieve complete surprise and take the maximum advantage of long-range missile attack. Such a long-range attack would be made using radar-guided missiles, shorter-range by heat-seeking and close air combat by guns. But the microprocessor has revolutionised air-to-air weapons just as it has the rest of the air defence environment, and all-weather, all-aspect, missiles have blurred the traditional scenarios for the use of the three kinds of weapon.

Radar-guided missiles. A modern radar-guided air-to-air missile depends initially on the illumination of the target by the air intercept radar in the nose of the fighter. In the closing stages of the attack, the weapon is homed on to the target by the missile's own radar guidance system. Target destruction is achieved by either impact or by detonation of the warhead by a proximity fuse. One of the most advanced is the AIM-54C Phoenix produced by the Hughes Aircraft Company. Phoenix has a range of over 100 nautical miles, a speed in excess of 3,000 mph and carries an expanding rod warhead weighing 133 lb. Controlled by the AWG-9 fire control system in a USN F-14 Tomcat, six targets can be engaged simultaneously. In a multiple engagement the missiles will attack on internal guidance from the missile radar after initial target illumination by the interceptor. At short range for self-defence the missile can be launched at a target without any assistance from the parent aircraft. The AIM-54C has a new digital electronics unit, inertial navigation reference system and a solid state radar transmitter, which together with sensitive flight controls enable it to engage highly manoeuvrable small targets, including cruise missiles, at low level. The combination of the AWG-9 aircraft radar and the missile's own characteristics comprise what is usually referred to as a 'look-down, shoot-down' capability against aircraft hitherto protected from radar-guided missile attack by the surrounding 'clutter' emanating from ground returns. The British Tornado F Mk2 will carry British Aerospace Sky Flash missiles which can be launched against high- or low-flying targets using a monopulse Doppler radar which can discriminate between individual targets, distinguish returns from low-flying aircraft and increase immunity to electronic counter-measures. The great advantage of a missile which can assume autonomous target acquisition is that the interceptor can break off its own attack either to seek further targets or to manoeuvre for its own security. A major factor in their effectiveness, however, must be the certainty that targets attacked beyond visual identification range are in fact hostile: hence the considerable importance of

PLATE 2.9. The AIM-120 AMRAAM being examined before despatch for test launch by the USAF (Hughes Aircraft Company).

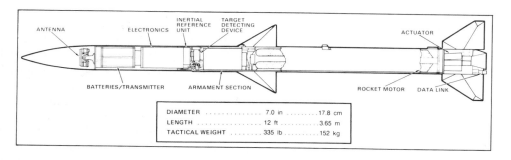

FIG. 2.4. AMRAAM (Advanced Medium Range Air-to-Air Missile).

accurate early warning information. A new long-range missile under development for NATO by Hughes, known as AMRAAM (advanced medium range air-to-air missile), will benefit still further from microminiaturised solid state electronics (see Plate 2.9). It will be two-thirds of the weight of the AIM-7 Sparrow which it is replacing, it will be guided by its own inertial reference unit, updated from the interceptor's own radar system and receive terminal guidance from its own active seeker (see Fig. 2.4). The interceptor pilot will thus have a complete 'launch and leave' capability.

Heat-seeking missiles. The second family of air-to-air guided missiles acquire their targets by using passive infrared (IR) sensors to home onto the heat generated by any aircraft in flight. The most commonly used in the West is the AIM-9 Sidewinder, in service with many air forces. Early models were effective only in attacks from astern, to reach the heat sources of the target's engines, but the missile has been progressively refined with increasingly sensitive heat-seeker and highly manoeuvrable shape to the point where the modern AIM-9L can be used at short range in all-aspect attacks while its successor, the AIM-9M, will discriminate between the target and infrared counter measures. Hitherto, flare decoys and even background sunlight as well as cloud and bad weather have attenuated the heat-seeking missile's effectiveness, but future generations will probably combine different kinds of guidance before final IR target acquisition. The French Matra Magic 2 is already reported to have a 'homing capacity on a designated target' in addition to the standard passive infrared with autonomous search. Future development facilitated yet again by the microprocessor may permit missile launch on inertial guidance, update on target location by data link from the interceptor's fire control radar and terminal acquisition by either IR or autonomous active radar.

Guns. When the massed fire power available to the interceptor is added up—the F-15 Eagle and the F2 Tornado, for example, both fly with four radar-guided and four IR missiles—it may be thought that the traditional gun, lacking sophisticated microcomputerised technology, has no place in the interceptor's armoury. Yet F-15 Tornado F. Mk2, F-14, F-16, F-18 Mirage 2000 and all recent Soviet fighters are equipped with various calibre of cannon. There are several reasons. Even modern all-aspect missiles such as AIM-9M have limitations on their use at low level in tightly manoeuvring combat and, as with any military technology, an offensive development very quickly spurs a defensive counter. Just as flare decoys and IR frequency jamming impaired earlier IR missile effectiveness, counter-measures will be devised, no doubt also drawing upon solid state electronics. Similarly, counters to missiles such as Phoenix or AMRAAM will emerge. Gunfire, on the other hand, lays the premium on pilot skill and aircraft maneouvrability and can be sustained by relatively inexpensive ammunition in large quantities; it cannot be diverted by decoys or electronic interference. Finally, it provides a valuable supplement in multi-aircraft combat to an interceptor which has launched all its guided missiles. On several occasions in the past, enthusiastic protagonists of the missile have forecast the obsolescence of the cannon. Most recently, the very high missile kill rate of the Israeli Air Force over the Bekaa Valley in 1982 was cited as evidence for the superiority of the IR missile. It may well be, but it would not necessarily be wise to draw infinite conclusions from a concentrated campaign fought in cloudless skies in which one side enjoyed absolute technological and aircrew superiority, tactical surprise and immunity from an offensive or disruptive response. Perhaps of longer-term significance was the lesson learned by the USAF and USN in South-East Asia when it was found necessary to rediscover the ancient art of combat manoeuvring with both guns and missiles.

SURFACE-TO-AIR DEFENCES (SAD)

Strictly speaking, SAD do not comprise part of air power, but because of their impact upon it, as a major contributor to both strategic and tactical air defence, their deployment and associated technology require analysis.

SAD comprise surface-to-air guided missiles (SAM) and anti-aircraft guns (AA). Missiles, like their air-delivered counterparts, can be guided by radar, by infrared sensors, or optically. Guns can be either free firing, using optical sighting, or be associated with radar for target detection, acquisition and aiming.

Surface-to-Air Missiles

SAM designed for strategic or large area territorial defence have traditionally been designed to combat medium- or high-flying targets, although modern technology now produces systems such as the US Patriot which can be deployed to counter low-flying aircraft also. The missile components are the propulsion unit, the guidance mechanism and the warhead. The missile itself will usually be only one part of a complex system which will include target warning, identification and communication units. Ground launching frees the missile from the size and weight constraints placed upon its airborne counterpart, permitting a much larger propellant and warhead but conversely demanding much greater power to lift and sustain the missile's flight upwards towards its target. Because the target will usually be picked out against the sky, there are few of the problems associated with the need in a 'look-down' missile to acquire its target against ground clutter. The Russian SA-2 missile is probably the most well known of its kind, achieving instant publicity when, at the thirteenth attempt, one brought down a US U-2 in 1960. The weapon was used extensively in combat in South-East Asia and the Middle East and is in service with twenty countries. It weighs approximately 5,000 lb, has an effective ceiling of some 80,000 ft, a slant range of 31 miles and a conventional high-explosive warhead of 288 lb. It is guided to its target by automatic radio commands passed from its associated ground-based early warning, height-finding and acquisition radars. Later models may have on-board homing sensors. Its solid propellant booster motor and liquid propellant sustainer give it a speed in the region of Mach 3.5. A heavier and longer-range weapon, the SA-5, has also been exported by the Soviet Union, although by 1985 only to Libya and Syria. SA-5 is believed to have a slant range of 185 miles and an effective ceiling of over 90,000 ft, carrying either a nuclear or a conventional warhead. Like SA-2, it has depended entirely on external radar guidance hitherto, but later models may have been fitted with a passive anti-radiation seeker. If so, it may have been designed to threaten AEW aircraft operating beyond its own immediate airspace. In any event, however, both SA-2 and SA-5 are generally believed to be obsolescent weapons, readily detectable from the air because of their smoke trail, easy to avoid because of their lack of maneouvrability, and relatively easy to jam. Not so in the case of Patriot, which can search for, acquire and attack targets from very low level up to 80,000 ft with a slant range of 30 miles. It is believed to have a speed of Mach 5 and is guided by a combination of command and semi-active homing. Solid state technology passes target information by data link from the missile's own seeker back down via a ground computer so that command guidance can be instantly updated. Its high speed, manoeuvrability, resistance to jamming and fire target multiple engagement capability per battery make it a formidable replacement for Hawk and Nike missile systems previously protecting Western Europe. Friendly aircrew are likely to welcome especially its greatly improved IFF system. Earlier generations of SAM have either no or very rudimentary IFF systems, as the Mig pilot trying to shoot down the same U-2 in 1960 found to his cost.

Patriot also illustrates the tendency among SAM to be optimised towards the faster, low-flying and probably manoeuvring aircraft and it is in the area of tactical air defence, the subject of the following chapter, that the most marked developments in SAD are taking place. Nevertheless, the contribution of systems like SA-2 and SA-5 to modern air warfare remains considerable. First, they forced aircraft such as B-52 and the RAF's V Force to drop down to low-level operations, thereby forcing them within range of smaller but more numerous and equally threatening defences as well as reducing their unrefuelled operational range. Second, they increased the vulnerability of any large aircraft with restricted manoeuvrability such as transport, maritime or strategic reconnaissance aircraft, with the exception of the high-speed, heavily self-protected SR-71. Third, they force circumspection and the need for protection upon AEW aircraft such as E-3, Nimrod or E-2C. Even well-proven systems such as the British Bloodhound, at one stage transferred from the United Kingdom to provide protection in overseas areas, can close down an option or severely complicate offensive organisation. As a result, Bloodhound has now been redeployed to supplement the manned interceptor and shorter-range systems in the United Kingdom itself.

THE AGGREGATE

The provision of territorial or strategic air defence in the West has, since the emergence of a long-range Soviet bomber threat in the last generation, been extensively modernised. The means have been provided by the microprocessor and its associated solidstate microcomputers. Target information has been acquired earlier, more accurately and more comprehensively; communications have become more secure, multi-faceted and integrated; weapon guidance systems have become much more accurate, discriminating and jam-resistant; weapons have become lighter with high kill probabilities over longer ranges. In the struggle for tactical air superiority, the traditional arena of 'the fighter', developments have been no less dramatic.

3

Air Defence: (2) Air Superiority and Counter-air Operations

With the increasing range of multi-role aircraft, the traditional distinction between territorial or 'strategic' air defence and the denial of a particular area of airspace to an enemy is becoming blurred. For example, in a conflict in Central Europe, the integrity of West German airspace would be synonymous with control of the air over a theatre of warfare. RAF F4 Phantoms in Germany could be called upon to intercept hostile aircraft *en route* to targets near, or even across, the North Sea, as well as to participate in the struggle for air superiority in the immediate vicinity of a land offensive coming across the inner German border. Similarly, in the Middle East conflicts of the last twenty-five years, fighters on all sides have been required to discharge both roles. Syria has been supplied with the Mig-23 by the Soviet Union while Israel has used both F-4 Phantom and F-15 Eagle.

The contest for regional air superiority is not decided solely by decisive fighter action. As in strategic air defence, the surface-to-air missile and gun have a major role to play. One additional measure is, however, associated far more with air superiority than with strategic air defence, and that is the counter-air operation. Counter-air is the name given to an offensive tactic which has the objective of contributing to air superiority by destroying or neutralising enemy air assets either in the air or while still on their home base. Even with the advent of the longer-range fighter bomber in both East and West, the bulk of any close air support to ground forces will be provided by aircraft based within range of the opposition's own fighter bombers. From the earliest days of air warfare, the destruction of enemy aircraft on their own airfields has been a primary role of tactical air power. It remains so in the present decade.

THE AIR SUPERIORITY FIGHTER

The air superiority fighter (see Plates 3.1, 3.2 and 3.3), or as it is increasingly labelled the 'tactical' fighter, is the prima donna of air warfare and even in the age of stand-off weaponry an essential element in all modern air forces. Its primary role is to combat either similar hostile aircraft in the contest for local air superiority, or to intercept and destroy hostile fighter bombers or bombers in the immediate vicinity of warfare on sea or land. It may in addition be called upon to

39

PLATE 3.1. The British Experimental Aircraft Programme Demonstrator. A single-seat air superiority aircraft with all moving foreplanes, advanced compound sweep wing and powered by two Turbo-Union RB Mk 104 engines. These features, together with new lightweight materials and active control technology, result in an outstandingly agile aircraft. New developments in stealth technology are also incorporated (British Aerospace).

contribute directly to a land campaign by attacking targets among enemy ground forces. The airframe must be highly manoeuvrable, able to accelerate rapidly, sustain high rates of turn and climb and destroy targets at short and medium range, ideally at night and in all weathers. Its pilot must possess very quick reflexes, excellent eyesight, physical fitness to maintain combat effectiveness under sustained acceleration forces and be able to communicate swiftly with colleagues in the air and on the ground. In addition, he must receive early warning of threats from air-to-air and surface-to-air weapons and be given a degree of protection from them.

As was summarised in Chapter 1, manoeuvrability or agility is conferred by aerodynamic design and construction, instantaneous controls and high thrust-to-weight ratio propulsion which gives spare energy over and above that required for straight and level flight. Among contemporary tactical fighters, the F-16 and F-15 (see Plate 3.4) are generally recognised to be clearly superior. The F-16, originally an experimental aircraft, is comparatively small with a wing span of 32 ft 10 in, a length of 49 ft 5.9 in and empty weight of 15,586 lb. The F-4 Phantom, which it was designed to replace, had comparative statistics of 38 ft $7\frac{1}{2}$ in, 63 ft and 30,328 lb. Both aircraft can carry a wide range of offensive and defensive equipment and both are capable of speeds in the region of Mach 2. The F-16 drew heavily upon the emerging technology of the 1970s: increased use of composite materials in airframe construction, fly-by-wire controls, blended wing–body aerodynamic shape, automatic wing leading edge flaps and a single Pratt and Whitney turbofan engine which gave 25,000 lb of thrust with afterburning giving a 1.1 to 1 thrust-to-weight ratio. The pilot's ability to operate the aircraft was facilitated by a high *g*

PLATE 3.2. F-18 air superiority/ground attack aircraft (McDonnell Douglas).

tolerance cockpit with a 30° reclining seat, high-visibility bubble canopy and head-up instrument displays. The result was a fighter which had a relatively large wing area, inducing low overall wing loading, but with the size of the wing more than compensated for by other technological advantages. Its rate of climb was second only to the F-15, while its turn rate at Mach 0.9 of 17° per second was superior to any other fighter. F-16 is not only important because of its contemporary superiority, but because future fighter designs appear to be using its capabilities as a starting point.

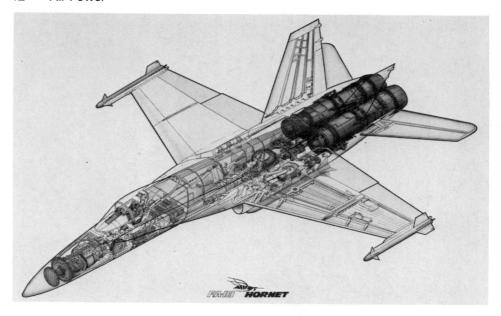

PLATE 3.3. Major components of the F-18 (McDonnell Douglas).

PLATE 3.4. F-15C with typical multi-role payload of air-to-air missiles and freefall bombs
(McDonnell Douglas).

Future Developments

Already the technological opportunities in airframe design are taking shape. The aeroelastic construction of the Grumman X-29A is one example; Boeing's Mission Adaptive Wing (MAW) is another, whereby improved turn rates can be achieved not by increased thrust or increased wing area, but by computer controlled reshaping of the wings in flight. The MAW has no flaps, spoilers or other hinged surfaces and is not only lighter than a traditionally constructed equivalent but will present a lower radar signature. Other proposals embody the use of 'canards' or small wings forward of the mainframe, as in the Swedish JAS 39 Gripen or Israeli Kfir, to enhance manoeuvrability. A typical fighter specification published in 1985 identified the need for a sustained turn rate of 20° per second at sea level and a maximum attainable rate of 30° per second with engine power designed to maximise acceleration rather than simply high speed. Digital engine controls will enable fuel consumption to be reduced even at supersonic speeds, thereby either reducing overall aircraft fuel requirement and associated weight or releasing spare capacity for other avionic or weapon options. Several projects incorporate two-dimensional thrust vectoring engine nozzles to improve short take-off and landing characteristics without incurring the design complexities of a supersonic VSTOL or STOVL fighter. The application of modern engine internal aerodynamics to reduce the number of compressor stages will increase effectiveness, reduce weight and simplify engine maintenance and support. A single seat is desirable to reduce overall weight and size still further, but at a cost of increasing pilot loading.

Cockpit Requirements

As a result, considerable emphasis is being placed on cockpit design and what is referred to as the 'man–machine interface' or MMI. The additional demands on the fighter pilot of the future will stem from three developments: the impact of high g loading in increased turn and acceleration rates; the amount of information presented to him from digitalised instruments and sensors; and the closing and combat speeds induced by improved airframe and engines. To reduce g forces the inclined seat as in the F-16 will become the norm, with manual controls positioned so as to reduce the physical effort required to operate them; for example, the placing of the control column to the side, rather than in front, of the pilot and close proximity fingertip selector buttons.

Displays

Information will be presented on demand, rather than permanently, in the pilot's line of vision rather than below the cockpit canopy in traditional layout. At present, most information is available to the pilot on a large number of instruments placed below his eye level, forcing him to look down inside the cockpit to monitor them. A certain amount of data is already projected onto a head-up display (HUD) by a cathode ray tube (CRT) generated image onto a semi-transparent mirror located in front of the pilot level with the cockpit canopy. The image, projected by an optical system known as a collimator, is focused to infinity so that the pilot can see both the outside view and the data at the same time without either having to move his head or refocus his eyes. The data is presented in the form of symbols which provide him with the basic parameters necessary for him to fly and fight: attitude, heading, altitude, speed and target-sighting marks. However, not only is the HUD information limited, but the HUD itself is very small compared to the area outside the cockpit which the pilot must habitually scan. Plates 3.5 and 3.6 show present and future cockpit visualisation systems.

Now the application of diffractive or holographic optics allows the construction of a larger display, increasing the field of view from 15 degrees in azimuth and 18 in elevation to 30 and 30. Moreover, the larger display area can be used to project the greater volume of information available from a digitalised, integrated systems unit and presented either automatically or on demand by the pilot. The contribution of the computer is to prepare information for display from sensors, process it and present it in a form most easily assimilated by the pilot. The same integration system can be used to display further information immediately below the HUD either at or only marginally below the pilot's eye level. Similar optical techniques can project either images or alphanumerical data at infinity, so that again the pilot does not need to refocus. Finally, but demanding even more complex technology, target information can be projected onto the helmet visor immediately in front of the pilot's eyes. The helmet-mounted sight (HMS) display could in theory be integrated with the other associated information systems in the cockpit and would considerably enhance the pilot's ability to acquire and attack 'off-bore sight' targets, i.e. not in line with the aircraft's latitudinal axis. In highly manoeuvring combat, using all aspect weapons such as AIM-9L and M, it would be the ultimate refinement in perpetually aligned HUD and outside view. However, the weight of the equipment would be limited not only by normal physical neck loading but the considerable increase under high g, while the exact relative location of the sight to the sensor would not, as in the case of fixed cockpit displays, remain constant. Nevertheless, the prospect of greatly enhanced target acquisition is driving, especially in the United States, considerable research on the HMS.

Pilot Reaction

But no matter how agile the aircraft, or how much information the pilot receives, if he cannot make use of it the investment is worthless. Increasingly, the combination of aerodynamics sensors and automated controls will free the pilot from concern about the operating parameters of his aircraft and leave him more opportunity to concentrate on the assimilation of and response to target and threat-related data. Already he is made aware of a target's position within a missile 'envelope' and of his own vulnerability to attack by radar-assisted weapons. The incidence of erroneous response can be reduced by electronic locks on switches and other controls, as well as the mechanical 'safety' or positive unlocking devices associated with traditional instruments and controls. But it is speed of positive response which needs to be facilitated rather than inhibitions, and to this end research is taking place in several countries into speech communication between pilot and instruments. The interaction of developments in speech analysis and synthesis, together with the opportunity to use miniaturised equipment, has led to

PLATE 3.5. The Mirage 2000 cockpit, fitted with an integrated head-up/head-down Thomson-CSF display system (Thomson-CSF).

PLATE 3.6. Mock-up of a cockpit visualisation system developed by Thomson-CSF for future combat aircraft, including holographic head-up display and colour CRTs (Thomson-CSF).

experiments with systems designed either to respond to the pilot's voice or to transmit information to him orally. Some observers are sceptical about the prospect for voice commands because of the likely circumstances of air-to-air combat: extraneous cockpit and aircraft noise, g loading and other audible signals such as those from the radar warning receiver. Nevertheless, in the continuing search to maximise the effectiveness of human judgement and high-volume computerised information to produce rapid response, it is likely that voice command systems will at least be tested in future experimental fighters.

Weapon Systems

The armament of the air superiority fighter is drawn from the same stock as that of the interceptor described in the previous chapter. In a tactical environment, the combat is likely to be more often at close quarters and therefore a tactical fighter could be unlikely to take off armed only with longer-range radar-guided weapons such as Phoenix or Skyflash. Obviously such weapons would allow a long-range opening shot or shots in a multi-aircraft engagement, but thereafter the combat could be decided by a short-range, highly agile all-aspect attack missile such as the AIM-9M Sidewinder, or by the aircraft's cannon. Considerable professional debate has taken place in both East and West during the last decade about the likelihood or otherwise of sustained manoeuvring combat in an age of increasingly effective air-to-air guided missiles. The consensus is, and not just among fighter pilots, that manoeuvring combat will remain an integral element in the contest for air superiority: hence the considerable investment in both fighter aircraft and air-to-air weapons. At the time of writing, comments on laser air-to-air weapons below the level of outer space were restricted to speculation about highly classified experiments. Problems of power generation, not yet reduced by the microprocessor, appeared to preclude early installation in combat aircraft, but neither East nor West was disposed to draw attention to any possible research taking place in that area. Open press reports on future fighter evolution all assumed the delivery of conventional warheads, varying only in their type of fusing and explosive charge.

COUNTER-AIR OPERATIONS

Conventional warheads of a very different kind also play a major part in the second element in the contest for air superiority: the destruction or disruption of enemy aircraft on their own bases. 'Counter-air' operations are as old as air warfare itself: destruction of enemy airfields was widely practised in both world wars, with a particularly spectacular success achieved by the Luftwaffe against the Western allies on New Year's Day in 1945. The most devastating and decisive attack was that by the Israeli Air Force against her Arab neighbours at the outset of the June war in 1967, when rows of Egyptian, Iraqi, Syrian and Jordani aircraft were destroyed on their own tarmacs. Indeed, the attack was so successful, conferring complete regional air supremacy on the IAF at a stroke, that analysts are generally agreed that no country is likely to offer a similar opportunity to an opponent again. And yet to allow an enemy unimpeded use of his airfields, from which he could fly to participate in an air–land battle, would be to grant him an unacceptable and possibly decisive advantage. For example, in Central Europe, forty-four Warsaw Pact main operating bases were reported in 1985 to lie within 300 km of the inner German border, with another twenty-eight located within 800 km. In addition, many other subsidiary airfields capable of mounting fighter or fighter-bomber operations were situated in a similar radius.

In the traditional manner of warfare, technology has both facilitated attacks against such targets and at the same time made them more problematical. For example, since the bitter Arab experience of 1967, both NATO and the Warsaw Pact have 'hardened' almost all their main operating bases within range of the opposition's tactical aircraft. That is, hardened shelters have been provided for aircraft dispersal, control centres, fuel and weapon installations and crew quarters. A casual visitor to RAF Bruggen, for example, could be forgiven for believing that the airfield was deserted, except for one or two Tornados actually

PLATE 3.7. German Air Force Tornado in protective hardened shelter.

preparing to take off or recover. All the others would be inside the hardened shelters behind heavily reinforced doors (see Plate 3.7). The shelters themselves are usually placed haphazardly around the airfield, with doors facing in different directions. Thus a counter-air operation aimed in traditional manner against the aircraft themselves would offer little chance of success unless mounted to coincide with a mass take-off or recovery of several aircraft: not an impossibility, but extremely difficult to coordinate and time, even with real-time accurate intelligence. Otherwise a single, heavy, concrete-piercing weapon would be required for each shelter, and even then with no guarantee that the shelter was occupied at the time of the attack. Multiple passes, a heavy weapon load, pinpoint accuracy with each weapon and, above all, invulnerability to airfield surface-to-air defences would be the total requirement. Not surprisingly, neither NATO nor Warsaw Pact forces have apparently sought to pursue such an option. A second possibility would be to attack weapon and fuel stocks on the bases. But these are usually widely dispersed, probably underground, certainly hardened and possibly well hidden. Good intelligence could probably make such targets vulnerable to precision-guided weapons, but again pinpoint accuracy would be required against widely dispersed targets with no guarantee that fuelled and armed aircraft in the shelters would be disrupted. Technology in counter-air operations has therefore been increasingly concentrated on the operating surfaces of the airfields themselves.

Recent generations of jet fighters and bombers have tended to require 2,000 yd or more of reinforced concrete runway to get airborne with full weapon loads. If that were still the case, airfield interdiction would be relatively easy to accomplish, airfield defences apart, by little more than one or two cratering attacks. Now most main bases have at least two runways, plus perimeter tracks which are regularly used for take-offs and landings. Several Warsaw Pact aircraft have been built with rugged undercarriages to facilitate operations from firm earth strips and increasingly modern combat aircraft are designed from the outset to operate from 500 yd of runway or less. The RAF Harrier and the USMC AV-8B can either take off and land vertically, or use very short distances, while most advanced fighter designs incorporate some form of short take-off and landing capability. Therefore, even airfield denial operations involve complex technical problems, most of which are being met in a broadly similar manner in East and West.

Air-to-surface weapon accuracy, as explained fully in Chapter 4, has increased considerably during the last two decades. Weapons to be used against runways require very specific characteristics. Local surface-to-air defences are likely to force the attacking aircraft to approach at high speed at heights probably less than 200 ft. Assuming, as is the case at present, that the aircraft must overfly the runway, three practical problems must be overcome. First, the aircraft must survive the effects of blast and debris from its own weapons; second, the angle of weapon attack must be sufficient to eliminate possibilities of ricochet; and third, explosive detonation must crater the runway sufficiently to make repair time-consuming and disruption extensive. The most widely used specialist anti-runway weapon in service in 1985 was the French Matra Durandal bomb. It could be dropped from 250 ft at aircraft speeds up to 600 knots. On release, it is slowed by a small brake parachute, then stabilised by a second one into an angle of descent of 30–40°. A few feet above the ground a rocket motor is ignited which drives the bomb at high speed into the runway. One second after penetration the 100-kg high-explosive warhead is detonated, reportedly penetrating concrete up to 40 cm thick and destroying a total runway surface area of 150–200 m^2. The technology involved in Durandal is relatively simple: standard delayed action and timed mechanisms which are preset, and once the weapons are airborne the procedures are automatic.

Another Western anti-airfield weapon is electronically operated and, unlike the unitary Durandal, uses advanced submunitions. The British JP233, built by Hunting Engineering, combines two sets of weapons in one ventral pod container on the attacking aircraft. Thirty of the submunitions initially, like Durandal, stabilised by parachute, attack the runway, but then a distance sensor detonates a shaped charge which pierces a hole in the runway; the second HE warhead passes through the hole and is detonated below the surface by a time fuse. Simultaneously with the launch of the anti-runway weapons, 215 area denial, delayed action minelets are scattered across the runway and surrounding surfaces. Release of both sets of submunitions is achieved by a computerised electronic distributor which generates a sequence of pulses preprogrammed with aircraft type, speed, height and attack heading to ensure even distribution. While the one group of weapons penetrate the runway, the others are programmed individually with a wide range of delays and intermittent fusing sensitivity. Each mine contains a

PLATE 3.8. The JP233 airfield attack weapon system in dual tandem dispenser on a Tornado.

PLATE 3.9. RAF Tornado releasing JP233 runway cratering and area submunitions during trials (Hunting Engineering Ltd).

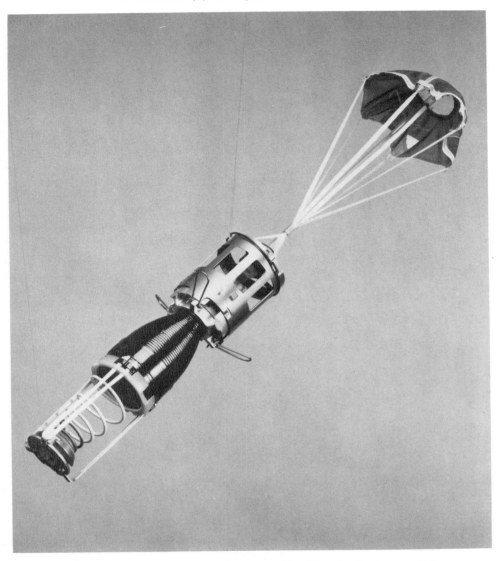

PLATE 3.10. JP233 runway cratering submunition (Hunting Engineering Ltd).

solid high-velocity slug powerful enough to pass through a bulldozer blade and destroy an engine beyond it, while on explosion the casing disintegrates into self-forging fragments effective over large areas against aircraft and personnel. In this weapon, the microprocessor and advanced submunition technology combine to threaten both airfield operations and subsequent repair. Plates 3.8–3.12 show JP233 sequence.

In the near future, other technological developments will be incorporated into the anti-airfield operation. Surface-to-air defences, concentrated near a known target such as an airfield, can be expected to take a heavier toll on the fastest and

PLATE 3.11. JP233 area denial minelets among the debris of a damaged runway (Hunting Engineering Ltd).

PLATE 3.12. Impact of JP233 runway cratering weapon, illustrating crater and 'lifted' surrounding area (Hunting Engineering Ltd).

lowest of attacks, even with on-board protection. Consequently, increasing investment is being made in a combination of stand-off guided missile systems and submunition performance. The stand-off weapon container could be launched by aircraft outside terminal SAD range, guided or preprogrammed by digital systems which could, if necessary, be updated after launch by data link. The submunitions would be either runway penetrators or, as with JP233, a combination of surface destruction and area denial. In any event, the defensive capability of even the most sophisticated of point SADs will continue to be stretched to the full. Indeed, no matter how complex airfield defensive systems become, the offensive drive to penetrate them will be sustained; the penalties to be imposed, and the advantages to be gained, are simply too great to be abandoned.

Even in an age when single weapons have a far higher kill probability than at any previous time, or when submunitions and warhead technology are increasing the lethal area of an attack, there are many occasions when the precise timing of an attack may be crucial to its success. For example, if air support is to be closely integrated with an attack on the ground, either as a preliminary bombardment or concurrent close air support; or if a beleaguered ground force calls for speedy air support; or if hostile air attack on ground forces is expected; or if an element in airspace management is the allocation of time periods for the unimpeded passage of friendly aircraft: in all these instances and many more, swift response and/or adherence to planned timing would be critical. Some air superiority fighters, such as the F-15 and F-16, have the endurance to take off and loiter for a limited period before being directed by a ground controller either to an air-to-air contest or to contribute to the air-land battle. With in-flight refuelling (IFR) such endurance can be extended, but for the foreseeable future tactical fighters will frequently need to return to base to refuel and regardless of IFR to rearm. In that environment, two scenarios can be envisaged. In the first, there is no hostile air threat to the airbase. Air and ground commanders know that their plans can proceed without concern about disruption. They can assess ground crew turn-round times for each aircraft, fuel and weapon consumption, battle damage repair capacity and pilot loading very much as the Israeli Air Force seems to have done in 1967 and 1973 when no Arab air attacks were made on IAF bases in Israel. The dominant considerations affecting air operations on airfields under such circumstances are those which generate the highest number of sorties in the shortest possible time when they are required.

In the second scenario the objective, to produce the required number of sorties from the airbase at the right time, remains the same. Now, however, the threat of air attack forces consideration of other priorities and the allocation of considerable additional resources. A heavy and successful air attack would by definition reduce the air effort which could be mounted for as long as it took to repair damage to infrastructure to allow the remaining aircraft to recommence operations, quite apart from the need to make good any aircraft losses. That factor would reduce the reliability of any air support envisaged from that airfield and complicate any ground offensive planning which was to depend upon it. But even a lighter, less successful, attack could interrupt operations: at the minimum by the duration of the attack itself and thereafter increasing in delay, and hence in disruption, of any planned or demanded aircraft support. Delay and dislocation of the planned air component threatens a degradation of timely concentration either directly in a temporary contest for air superiority over a battle area or indirectly by reducing a capability to mount a coordinated, sustained air-to-ground intervention in the ground battle itself.

Finally, even if an air attack is never actually mounted against an airbase, the defender is compelled to allocate a costly amount of resources to ensure continued operations which themselves induce further vulnerability. Modern aircraft require extensive ground engineering support, and even high resilience to battle damage and good serviceability do not preclude the need for refuelling and rearming. The defensive provision for main bases has already been noted, and if plans are laid to recover aircraft to other bases, or dispersed airfields, the problems are far from resolved. Dispersed airfields demand more manpower, alternative fuel and weapon stocks and increased investment in logistic support. All require further defences, and both NATO and the Warsaw Pact have not extended full hardening and air defences to secondary airfields which in theory are capable of accepting and returning front-line aircraft to combat. Western aerial surveillance is already well capable of detecting Warsaw Pact deployments to secondary airfields within range of counter-air operations, and such deployments would inevitably leave aircraft far more vulnerable while on the ground. It may confidently be predicted, therefore, that counter-air operations will continue to be a fundamental element in the contest for air supremacy.

Whether they will continue to be discharged by aircraft overflying the target airfields is much more open to doubt. One school of thought argues that to send manned aircraft against an utterly predictable, immovable large-scale target such as an airfield is a waste of the basic advantages of air power: its flexibility and responsiveness to rapid changing circumstances. Technology, it is argued, provides the alternative approach of ballistic missile armed with terminally guided conventional submunition warheads. Weapons based on the Trident C4 booster rocket and the Martin Marietta Pershing missile are already being studied. Propellant, guidance systems and submunitions already exist; there is little doubt that a ballistic counter-air weapon could be introduced in the next decade. Such a weapon could respond to real-time intelligence reports of enemy preparations for take-off or recovery and would, in a similar timescale at least, offer virtually no options for its interception. The problem with such an attack in a European scenario is the difficulty of launching any surface-to-surface missiles without giving the impression of a nuclear missile launch and prompting instant but really nuclear response. The problem of discrimination, as the surveillance effectiveness of both sides increases, does not seem insurmountable, but may be sufficient, together with the air force preference for the greater flexibility of the manned aircraft, to rule the option out. If so, the preferred solution will lie in the introduction of stand-off guided missile technology into the counter-air aircraft/weapons system.

TACTICAL SURFACE-TO-AIR WEAPONS

Tactical SADs, like their strategic counterparts contributing to territorial defence, are not elements in the application of air power. They do, however, inhibit tactical air operations, contribute to the quest for local air superiority and therefore influence the technological evolution of the aircraft themselves. Their capabilities must therefore be considered. Tactical SADs also comprise guns and surface-to-air missiles, but whereas in strategic air defence the latter will usually be permanently sited, heavy weight and long-range, tactical SAM are normally short-range, mobile and frequently light enough to be man portable. They may be able to rely on early warning from strategic or tactical airborne systems, but more often will be dependent on their own surveillance and target acquisition systems and must be able to react with considerable speed. Increasingly, their targets will be either fixed wing aircraft flying below 200 ft or helicopters using 'pop-up' attack tactics in which they may be visible above natural skylines for only a few seconds at a time. For example, an aircraft flying at 540 knots (1,000 km/hr) will travel 3 miles in 18 sec. At 200 ft or below, it would be difficult over most landscapes to spot it either visually or by radar from the ground beyond that distance. Yet, assuming it is not carrying stand-off missiles, it must be detected, identified and, if hostile, intercepted within that time. In passing, the significance under such circumstances of AEW and elevated ground radar units should be noted.

In most European war scenarios it is assumed that the battlefield will not be static and therefore neither will the battle for air superiority above it. All tactical SADs must therefore be sufficiently mobile to keep pace with either offence or defence. Increasingly, both guns and SAMs are being mounted on tracked chassis, as for example the Italian Madis 4×25 mm gun. Together, the four guns can fire at a rate of 2,000 rounds per minute. The days of visual sighting and firing of AA guns are long gone. Madis uses a low-light TV target tracker, a day/night sight with a laser rangefinder, automatic or manual tracking, a ballistic computer and a vehicle attitude sensor. A video compatible Forward Looking Infra Red (FLIR) sensor can be coupled to the sight for use in haze or bad weather. The manufacturers claim an average reaction time between target acquisition and firing of 6 sec.

If a hostile aircraft is aware that such a system is in the vicinity, and the pilot will receive no warning of optically or unguided attack by ballistic rounds, he may well be inhibited from maintaining a consistent flight path. Hence one reason for airframe manoeuvrability, high rates of acceleration and a weapon-aiming system which can be independent of the aircraft's line of flight. By acting as a deterrent, the rapid-firing low-level gun can drive an attack pilot into a flight pattern more favourable to the second element in the SAD: the SAM. The tactical SAM may rely on infrared sensors and line-of-sight aiming, as with the hand-held Soviet SA-7 or its Egyptian offshoot, the Sakr Eye. The IR SAM has compelled tactical fighters to carry the additional weight of heat-emitting decoys and, in future, computerised systems to disrupt the IR wavelengths themselves. In turn, the IR SAM will be refined to allow it to discriminate more effectively between target and decoy by using both IR and ultraviolet sensors or even by seeking the second hottest target acquired by the IR sensor. As with the AIM 9 series of air-to-air weapons, an all-aspect capability and increased missile agility will enhance considerably the threat even to the high-speed manoeuvring target. Complete convergance will occur in the Luftwaffe in 1986 when AIM-9s will be deployed in a short-range ground defence system fully integrated with target acquisition and identification automatically processed over data links. Optically guided SAMs have the advantage, like the gun, of giving no warning of acquisition to the target aircraft. Conversely,

Each ADATS fire unit either uses its own autonomous radar to detect low-flying aircraft or is netted to larger radar systems for command and control. Once a target is detected it is passed to the ADATS passive optical tracking system (television or FLIR), which is highly resistant to antiradiation missiles and electronic countermeasures. A supersonic missile is guided to the target with a spatially encoded CO_2 laser beam. A proximity fuze detonates the special dual-purpose warhead for airborne targets and an impact fuze detonates a conical-shaped charge for armor targets.

System Performance
● Intercept range: greater than 8 kilometers (km)
● Ceiling: 5000 metres
● Missile velocity: greater than Mach 3
● 8 missiles ready to fire
● 8 additional missiles ready for use, depending on the carrier

Search Radar
● Airspace surveillance
● Range: greater than 20 km
● Dual beam system
● Search during the move

Missile
● Length: approximately 2.05 m
● Diameter: 152 millimeters
● Weight: approximately 51 kilograms at launch

Dual-Purpose Warhead
● Weight: more than 12 kg
● Electro-optical proximity fuze
● Impact fuze
● Hollow charge and fragmentation effects

Canister
● Length: approximately 2.2 m
● Diameter: approximately 240 mm
● Weight: approximately 13 kg

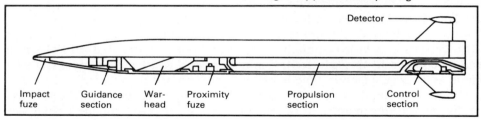

FIG. 3.1. Martin Marietta Air Defense/Anti-tank System (ADATS).

their effectiveness is degraded by battlefield smoke, bad weather and night. Consequently, greater emphasis is being placed either on supplementary radar guidance to what were originally optical SAMs, such as the British Rapier or French Roland, or the development of wholly radar-guided systems, drawing fully on solid state technology such as the Martin Marietta ADATS (see Fig. 3.1). As in other areas of air combat, a major contribution of the microprocessor is to allow considerably more information to be acquired, analysed, transmitted and, if required, to prompt automatic responses. Thereafter, improved rocket motor technology, producing missile speeds in excess of Mach 4, together with missile agility capable of 30g turns, combine to add substantially to the attacking pilot's considerations. Such factors drive still further the need for agility, on-board identification, threat warning and defence suppression systems, all susceptible to computer assistance, but all making the contest for air superiority far more than the traditional eyeball-to-eyeball fighter pilot contest. In the further future, there is little doubt that conventional weapons will be supplemented by high-energy lasers. By 1981. the USAF had successfully tested surface-to-air lasers and for several years similar Soviet experiments have been taking place at Saryshagan and elsewhere. High-energy lasers obviously present an instantaneous threat to either aircraft structures, avionic systems or the pilot himself. But the defence is unlikely to have a laser monopoly; indeed, one of the publicised USAF programmes in 1981 was the use of a carbon dioxide gas dynamic laser to destroy air-to-air missiles. For the foreseeable future, air superiority is likely to be contested by manned aircraft and surface-to-air weapons; either in the skies over, ahead of or behind the battlefields, or on the airfields themselves. And air superiority is not even an end in itself; it is the inseparable means to a successful outcome of combat on land or at sea wherever an opponent has the means to bring his own air power to bear on the conflict.

4

Air-to-Ground Operations

In any analysis of air operations, classification by roles is impeded by overlapping activities between them. We have seen, for example, how counter-air operations against airfields can contribute to the achievement of supremacy in the air. 'Air-to-ground operations', on the other hand, usually refers to offensive or defensive contributions from the air to the combat taking place on the earth below as distinct from attacks on airfields. Even then, the term needs to be refined still further because air power not only contributes fire power to the ground war, it also provides reconnaissance and air mobility, both of which are described in later chapters.

Terminology varies between alliances and between individual countries within the alliances, but generally speaking aircraft will contribute to the land battle in one of three ways. First, by attacking enemy ground forces which are actually engaged in combat with friendly ground forces. This activity is known as Close Air Support (CAS). Second, by attacking enemy forces which may be closing to join the ground battle in the immediate future. These air attacks would take place a relatively short distance behind the battle area, perhaps up to 50 miles, and would seek to have an indirect but speedy impact on the battle itself. These operations are known as Battlefield Air Interdiction, or BAI. Third, a traditional use of offensive air operations is the role of deeper interdiction, whereby reinforcement and resupply of the battle area is denied by deep-ranging attacks on reinforcement routes and resupply areas up to several hundred miles behind the front line. All three activities are designed to influence one particular battle area, or theatre of operations, and are usually cited as examples of the tactical use of air power. Although target arrays may be similar, whether engaged in combat, preparing to enter it, or still assembling and moving from the rear, the operational environment of each provides very different challenges and not surprisingly modern technology has contributed several different options for their resolution.

In all the anlysis which follows in this chapter, two very important assumptions have been made: first, that reconnaissance or other intelligence sources have provided up-to-date information about the whereabouts of the targets; and second, that enemy interceptions have been neutralised as a result of measures such as those described in Chapter 3.

PLATE 4.1. Italian Air Force Tornado interceptor/strike attack aircraft on a training sortie.

CLOSE AIR SUPPORT

The ideal close air support could quickly be described by any soldier. It may be preplanned as part of a coordinated land–air battle, or be an instant response either to a no notice call for assistance by air attack on a particularly stubborn enemy defensive position, or to a hasty plea for heavy fire power on an opponent whose strength in attack is threatening to overrun the friendly ground forces. In either case, heavy fire power will be speedily, accurately and successfully directed against the designated targets regardless of climatic conditions, by day or by night, so that the friendly ground forces can thereafter either continue their own offensive or regroup to counter-attack. Technology is bringing that ideal closer to reality, but the reality, even when surveyed cursorily, is daunting. A modern land battle between all but the poorest of Third World countries is likely to be a mobile engagement in which an important, if not predominant, role will be played by armoured vehicles: tanks, troop carriers and self-propelled artillery. Interspersed among them will be the surface-to-air missiles and other anti-aircraft weapons described in Chapter 3. Combat is likely to be sustained by day and night, with little or no intervention by weather except perhaps in extremely low temperatures or flooding. The combat area itself is likely to be obscured at least intermittently by smoke, and targets will not be easy to detect or identify. Their formation is likely to be haphazard and their movement not necessarily predictable. Finally, the most formidable combat unit, the main battle tank (MBT), is the most difficult to destroy and is deployed by the Warsaw Pact countries in very large numbers.

THE AIRCRAFT

The air forces of NATO and the Warsaw Pact have, perhaps not coincidentally, approached in very similar ways the problems prescribed by the need for CAS. Both have either failed to decide which is the most desirable complement of aircraft and weapon system, or alternatively they have decided that more than one combination is preferable. The aircraft are either fast jet aeroplanes, relying on speed to minimise vulnerability; or heavily armoured and highly manoeuvrable slower jets; or helicopters which can use terrain features for cover and ambush.

Harrier II

The British Aerospace/McDonnell Douglas Harrier II (see Plate 4.2) comprises and contains much of the technology associated with the fast jet approach to CAS. The aircraft is powered by one Rolls Royce Pegasus II Mk 105 vectored thrust engine giving a speed of Mach 0.88 for low-level dash and a typical combat radius of action carrying seven BL 755 cluster bombs in excess of 600 miles, or proportionately shorter if called upon to 'loiter' in the combat area before attacking. The unique feature of the Harrier family is the vectored thrust of the Pegasus engine which supplements the airframe's lift, enabling it to take off and land vertically or with very short ground runs. Thereby, Harrier can readily deploy forward to semi-prepared operating bases to provide swift reaction to calls for CAS; conversely, it can use damaged runways which would render a more traditional fixed-wing CAS aircraft non-operational. In flight, the engine exhaust nozzles can be swivelled to enhance considerably the aircraft's manoeuvrability either in air-to-air combat or in breaching surface-to-air defences. Graphite epoxy composite materials have been used in the construction of the wing, tailplane and forward fuselage, with consequent weight savings. Aerodynamic improvements have been achieved by larger trailing edge flaps, drooped ailerons, under gun pod strakes and leading edge wing root extensions, enhancing both general performance and specific turning rates. A computerised departure resistant system allows the pilot to take the maximum advantages of the combined engine and airframe components of manoeuvrability without fear of control loss. Information on optimum speeds and heights for flight profiles as well as point-to-point navigational data are, as on most modern combat aircraft, provided automatically by integrated data systems. All essential information is presented on a digital control panel directly under the head-up display. The USMC's Harrier, the AV-8B, can carry freefall bombs, cluster bombs, rockets, air-to-ground guided missiles, cannon and air-to-air missiles, as well as 300-gallon drop tanks. Both Harriers are equipped with the Hughes angle rate bombing system, which locks on to a target and automatically releases the weapons. The system is effective against both moving and static targets and used with the aircraft's inertial navigation system, has produced very accurate bombing in weapon trials, reducing miss distances to a few feet and increasing considerably the proportions of direct hits. Such precision is obviously extremely important in an environment where distances between friend and foe on the ground could be very small. Self-protection against SAM threats is provided by a variety of systems which are explained in detail in Chapter 5 below.

PLATE 4.2. Harrier GR5 and its weapon options (McDonnell Douglas).

Thunderbolt II and Frogfoot

Just as the Soviet Air Forces have developed fast jet CAS aircraft, although not yet with the all-round capability of Harriers, so they have produced the CAS specialist 'low and slow' Su-25 Frogfoot, which in many respects resembles the USAF's Fairchild Republic A-10A Thunderbolt II. Thunderbolt II appears to have originated in USAF reflections on the war in South-East Asia, which had highlighted the high cost of fast jet CAS operations and the difficulties, with the technology available at the time, of target acquisition and destruction in high-speed low-level attack. Technology was applied to Thunderbolt II, and later to Frogfoot, to produce an aircraft which would rely more on armour and manoeuvrability to survive in the air–ground battle, and deliver heavy fire power for longer periods to achieve high kill probabilities. Titanium rather than carbon composites have been used to supplement traditional aluminium alloys, while aerodynamic efficiency has been sacrificed in the placing of two high by-pass ratio turbofan engines high on the aft fuselage, where their infrared signatures are shielded by the wing, but their contribution to airframe drag must be considerable. The aircraft has a modern INS navigational system, but was not designed with the advanced avionics or computerised weapon systems of aircraft such as the F-16 or Harrier II. Its major weapons are a seven-barrelled 30-mm Gatling-type rotary cannon which fires armour-piercing spent uranium shells weighing slightly more than 2 lb, at a rate of either 2,100 or 4,200 rounds per minute, together with the Maverick air-to-ground missile described below. The Su-25 Frogfoot is of more conventional design, powered by two turbojets which give it a higher speed but shorter endurance. It also carries armour plate and like the A-10A is probably fitted with duplicated control and other systems to reduce vulnerability.

Perhaps more significantly, Su-25 and A-10A demonstrate the limitations on CAS when modern technology is deliberately not incorporated in the aircraft. Neither can acquire targets at night or in bad weather. Both have to attack visually on line-of-sight, although it is probable that Su-25 carries a laser range-finder and A-10A can stand off to launch its missiles. It may also be significant that proposals to build an all-weather, two-seater A-10A were shelved and the production was restricted to 721, a relatively small number by US standards. Meanwhile, Su-25 has entered service not, as one might expect, with the 'sharp end' of the Soviet Air Forces in Central Europe, but with the Czechoslovakian Air Force. Both aircraft are designed to operate from dispersed bases on secondary airfields and prepared strips; both may be seen as an intermediate provision between the fast jet on the one hand and the heavily armed helicopter on the other.

The Offensive Support (OS) Helicopter

In many respects the helicopter comes closest to meeting the soldiers' CAS ideal. It costs appreciably less than the fixed-wing counterpart to build and to operate. It demands shorter and less vigorous flying training. It can operate without the infrastructure of an airfield with conventional runways. It can deploy forward with ground forces; it can lie in ambush behind low-lying terrain features; it can launch and recover to extempore sites, thereby extending practical mission endurance. The Soviet Air Forces had by 1986 deployed several thousand and were preparing to introduce two more advanced types into service; the United States was deploying its own advanced CAS helicopter, the two-seater Hughes AH-64A Apache. The Apache is the world's first helicopter designed from the outset for day/night all-weather CAS operations, drawing heavily on both traditional and microprocessor technology. The central systems of the Apache are the Martin Marietta Target Acquisition and Data System (TADS) (see Fig. 4.1) which is linked to a colocated pilot's night vision sensor using laser and infrared methods to provide target acquisition and identification. The Apache can take off with a gross weight of 20,100 lb, enabling it to carry as heavy a weapon load as its fixed-wing contemporaries, and all weapons, usually guided missiles, gun and rockets, are coordinated by an on-board integrated fire control system. In March

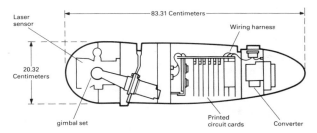

FIG. 4.1. Martin Marietta 'Pave Penny' laser target acquisition system.

1985 the fly-away unit cost was quoted at $7.03 m, making it less than half the cost of most of its fixed-wing competitors in the CAS role.

Yet despite its tactical advantages, and comparatively cheap costs, the OS helicopter was not regarded by Western air forces with unrestricted enthusiasm. Despite Apache's endurance of up to $2\frac{1}{2}$ hours with a full weapon load, there were those who believed that its limitations still precluded large-scale investment and the correlated reduction in fixed-wing provision which such investment might imply. Despite the addition of armour, mast-mounted sights to enhance conceal-ment in ambush, and operations at nap of the earth level, doubts about the helicopter's vulnerability to all types of ground fire remained. Transit and dash speeds remained well below fixed-wing capability, with attendant restriction on combat radius and increase in vulnerability. It was significant that the issue of helicopter vulnerability was addressed unequivocally by a German concept in 1986: the aircraft was unprotected against direct fire from missiles or guns. It was envisaged that it, the PAH-2, would attack tanks from ground cover, capitalise on surprise and then rapidly change position at very low level to avoid retribution. Critics of the concept pointed to a helicopter's vulnerability not just to ground fire but to other helicopters and, while refuelling or re-arming on the ground, to both artillery and air-to-surface attack. The cost penalty of dispersed operations, whether fixed wing or rotary, is the amount of logistic support required to sustain them. When forward operating bases are close to the front lines, the logistic supply lines themselves become lucrative targets for air attack and compound problems of comprehensive defence.

CLOSE AIR SUPPORT WEAPONS

Any distinction in offensive air-to-ground operations between CAS and interdiction weapons is artificial and incomplete, as most can be used in either role. When intervening in the land battle, however, the ability to distinguish clearly between friend and foe is of critical importance. Moreover, only a very short time may be spent in hostile air space, if indeed any at all. There is, therefore, less need to rely on stand-off weaponry to reduce vulnerability and every incentive to use accurate, if not necessarily terminally guided, weapons. Three kinds of weapon tend to be used under such circumstances: the ballistic rocket or gun; the laser designated bomb or missile; or the low-level, retarded but freefall bomb.

FIG. 4.2. The target-seeking concept using Ferranti marked target seeker in conjunction with ground designator.

With the increasing trend among the greater powers in recent years to transport infantry in armoured personnel carriers, opportunities for old fashioned strafing of exposed troops by machine gun have decreased. More typical of the modern air-to-surface gun is that already noted, firing 2-lb shells, carried by the USAF A-10A. The rocket, however, is still a popular weapon. A typical example is the Thomson Brandt SNEB rocket, 68 mm in diameter, 91 cm in length, fired singly or in salvos with an anti-armour shaped charge or general purpose fragmentation warhead. Ballistic rockets can possibly be evaded, but not jammed or diverted. They depend upon external target acquisition and, at present, line-of-sight aiming and against dispersed or manoeuvring targets they require very accurate shooting.

That accuracy can be considerably enhanced by relating a laser target designator and acquisition system with the weapon. Current equipment incorporates a laser target ranger and marked target-seeker as installed by Ferranti in the RAF Tornado GR Mk1 and Harrier. A laser on board the aircraft directs pulses of infrared energy at a target whose range is then measured by the time lapse between transmission and reception of the reflected energy. Alternatively, the target can be illuminated by an external laser source either on the ground or in the air (see Fig. 4.2). The marked target-seeker in the aircraft automatically detects the laser energy reflected from the target and is represented on the pilot's head-up display to indicate target position to him in elevation and azimuth. Such systems can be wholly integrated with aircraft avionics, not only to display target position to a pilot but to provide precise target location to weapon delivery computers, and in the immediate future to the rockets themselves after launch.

One such project is the Hughes Hypervelocity Missile Weapon System (HMWS) which incorporates a laser receiver and impulsive thruster control into a traditional rocket form, achieving a robust command guidance system for several missiles simultaneously. Without microprocessors such a weapon would be impossible; when operationally successful, it will combine the salvo fire power of the old fashioned rocket with the multiple-target accuracy offered by contemporary guidance systems. A similar principle can be applied in close air support, as in precision bombing elsewhere, when greater destructive impact is required. The RAF, for example, have given a new lease of life to old 1,000-lb freefall bombs by

PLATE 4.3. The mast-mounted sight enables helicopter gunners accurately to fire TOW anti-tank missiles from behind the cover of hills and trees (Hughes Aircraft Company)

'strapping on' laser guidance systems against illuminated targets. Laser designation under such circumstances allows the aircraft to break away from the area immediately after weapon release, even if the weapon is locked on to the target from the outset, when the weapon system has been self-illuminating.

If, however, close air support is being provided by helicopters rather than fixed-wing aircraft, a different family of weapons is available. As already explained, problems of target acquisition and identification, concealment and vulnerability call for different solutions. The Tube-launched, Optically-tracked, Wire-guided (TOW) anti-tank missile is carried by attack helicopters of many countries (see Plate 4.3). The helicopter gunner locates his target in his stabilised sight, fixes the target in the sight cross-hairs and fires. He then holds the cross-hairs on the target while guiding the missile to the target through two signal-steering wires attached to it. If the helicopter is fitted with a mast-head sight, it can remain almost wholly concealed by natural cover during the attack; if not, the stabilised system is unaffected by any manoeuvres undertaken. The basic optical system can be enhanced by laser range-finding and target acquisition and infrared amplification.

The Hughes AH-64A Apache attack helicopter will, on the other hand, be armed with the Rockwell International/Martin Marietta Hellfire laser-guided missile (see Fig. 4.3). Each missile carries a laser energy-seeker in the nose (compared with the tail-facing command guidance system in the Hughes HMWS) which locks onto energy reflected from targets illuminated by the Apache's own TADS designator or from an extraneous source. The Apache can carry up to sixteen missiles and can fire from ambush without disclosing its position to hostile air defence weapons as long as the target or targets are illuminated from another source or indirectly by the TADS. In ripple firing the missiles can be independently targetted using different laser codes from multiple designators.

Finally, among CAS weapons, the conventional freefall bomb has been adapted by more traditional technology for use against ground forces in contact with friendly troops. Targetting errors associated with freefall bombing have been produced by inaccurate aircraft navigation, imprecise bomb aiming and release and unpredictable weapon trajectories compounded by height and weather. The modern attack aircraft will be guided by inertial navigation systems, will have a variety of target range-finders, and will probably be joining a ground force battle area at low level. If it is carrying 'iron' bombs, it will probably adopt a lay-down technique whereby the weapons are dropped at very low level which, together with modern navigation and target acquisition systems, will reduce trajectory inaccuracies to an absolute minimum. The French Matra SFA is among the most modern of retardation systems, permitting the dropping of unguided bombs at heights of less than 100 ft at over 600 knots (see Fig. 4.4). Exposure to hostile ground fire and target inaccuracies are both reduced proportionately. Devoid of sophisticated solidstate guidance systems, the retarded freefall bomb is cheap to manufacture, can be carried by a variety of fixed wing aircraft and in units of up to 1,000 lb ensures the accurate delivery of heavy fire power. After release, the freefalling bomb is retarded by an integral parachute long enough for the aircraft to escape from the explosive blast only a very small distance below it. Lay-down bombing does, however, call for the attacking aircraft to overfly its target, albeit very fast and very low, and the system is arguably more suited to CAS, with maximum surprise for the defences, than for deeper penetration of hostile air space. Obviously, the deeper the penetration of that area behind the battle zone itself, the more complicated the offensive air support task.

AIR INTERDICTION

Interdiction of the battle area, whether close behind it or much deeper, calls for more extensive capabilities in the attacking aircraft than the direct contribution of CAS. Distances from home base are greater; the time spent within range of hostile defences will be much longer; the surface-to-air defences are likely to be more closely integrated than those caught up in the ground combat arena; the hostile early warning time will be longer; and there is less likelihood of the attacking aircraft being given last-minute target acquisition or identification assistance from the ground. As will be explained later, other kinds of help may be on hand from friendly electronic warfare aircraft, but, conversely, the deeper the penetration, the more scope for hostile defensive counter-measures.

The factors influencing a commander's decision whether to go for battlefield or deeper interdiction are complex, including an assessment of the timing and impact of any delay or disruption which he can achieve on the enemy's ground momentum. Even to permit him to make such a choice, his aircraft must be able to overcome the obstacles summarised in the preceding paragraph. In an aircraft

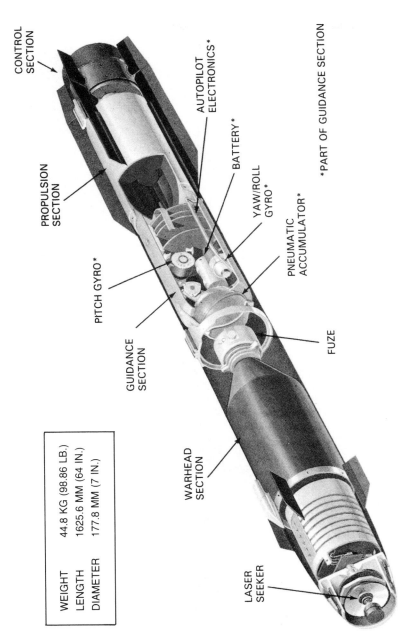

WEIGHT	44.8 KG (98.86 LB.)
LENGTH	1625.6 MM (64 IN.)
DIAMETER	177.8 MM (7 IN.)

CONTROL SECTION

PROPULSION SECTION

AUTOPILOT ELECTRONICS*

BATTERY*

YAW/ROLL GYRO*

PNEUMATIC ACCUMULATOR*

PITCH GYRO*

FUZE

GUIDANCE SECTION

WARHEAD SECTION

LASER SEEKER

*PART OF GUIDANCE SECTION

FIG. 4.3. The Hellfire laser-guided anti-tank missile.

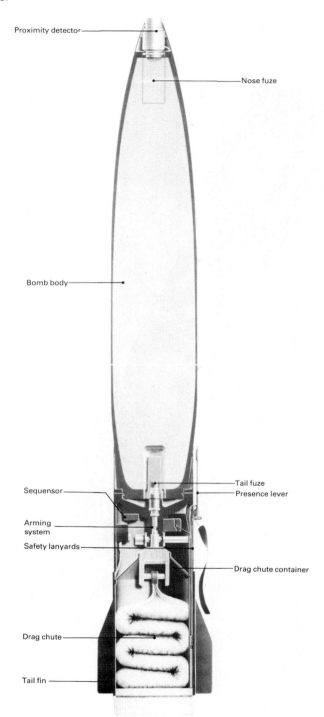

Fig. 4.4. Matra SFA retardation and arming system.

planned to enter service with the USAF in 1988, modern technology will address them all.

McDonnell Douglas F-15E Strike Eagle

The F-15E is the latest version of the F-15 air superiority fighter which first flew in 1972; only the shape is still recognisably the same. In its strike/interdiction configuration it will have a combat radius in excess of 1,750 miles, carrying 24,500 lb of weapons at high subsonic speeds at low level. When attacking targets well within that radius, range redundancy permits an indirect approach to reduce vulnerability to integrated air defences still further. The airframe has been strengthened to allow for increased take-off weight and uprated engines have been installed to give the heavier aircraft increased power, but it is again in the impact of the microprocessor that F-15E Strike Eagle is most notable. At the centre of the navigation and attack equipment is an IBM computer which receives data from a synthetic aperture radar and the LANTIRN (Low Altitude Navigation and Targetting Infra Red for Night) attack system. Information from the Synthetic Aperture Radar (SAR), married to the Inertial Navigation System (INS) track memory, is translated by the computer into a very-high-resolution map. At low level the computer adjusts the picture to remove distortions of angle and distance. Thereafter, the maps can be regularly updated or retained while the radar itself is switched off. The same radar can be used for long-range detection and tracking of aerial targets down to ground level, feeding data to the computer for fire control. Alphanumerical data is fed into the wide-angle HUD.

The second system, LANTIRN, comprises two auxiliary pods and a cockpit display (see Fig. 4.5). It has FLIR sensors (Forward Looking Infra Red), a laser target designator/range-finder, terrain-following radar and an automatic multi-mode target-tracker. The infrared imagery is projected on to the HUD, enabling the pilot to search and acquire small mobile targets on the ground at night and below cloud. Together the systems will allow the F-15E to operate round the clock in most weathers, as the radar will compensate when very low cloud or fog blacks out the optically based LANTIRN. In charge of the nav-attack systems of the F-15E will be a second crewman, the weapon systems operator (WSO), who will have four-colour CRT displays to help him operate the navigation, weapons, communications and defensive counter-measures systems. He also has a full set of flying controls should his pilot become incapacitated.

In sum, the conversion of what is basically a 1960s-designed aircraft into a 1980s all-weather strike attack weapon system has been achieved by the installation of equipment capitalising on the microprocessor and only supplemented by more powerful engines and a strengthened airframe. A similar transformation has taken place in the weapons which the latest F-15 version will carry.

INTERDICTION MISSION WEAPONS

In an interdiction attack, it may be assumed that all installations, armour, vehicles, SADs, etc, are hostile. Identification of friend or foe is not therefore the same problem and the value of high degrees of weapon accuracy and kill

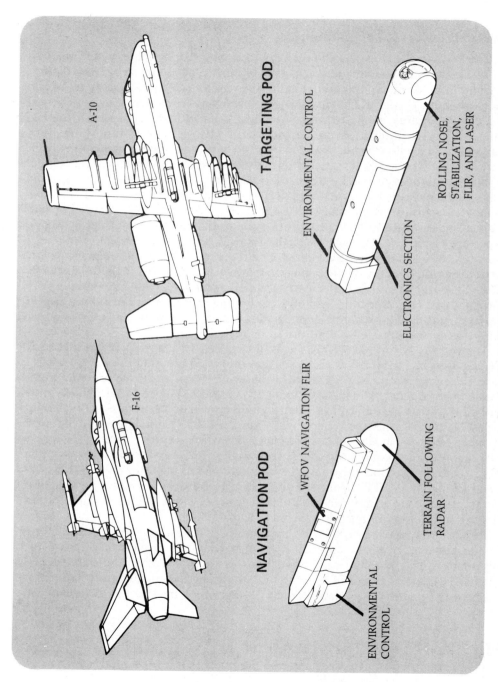

FIG. 4.5. The LANTIRN system.

PLATE 4.4. Canadian Air Force F/A-18 carrying Hunting Engineering cluster bombs on trials (Hunting Engineering Ltd).

probability lies solely in the exchange between aircraft and weapons launched against enemy assets destroyed. But as in CAS, while freefall bombs or rockets may be considerably cheaper per unit, their kill probabilities against either mobile or static targets are much less than PGMs, which also convey several other advantages. If the weapon has an effective homing mechanism, the weapon itself can be smaller, the logistic support and stockpile requirements are reduced, the number of sorties flown can be reduced and friendly casualty rates in turn diminish. Alternatively, the kill probability rate per weapon can be increased by making it a dispenser of guided or unguided submunitions. Three modern examples of such weapons are the Hughes AGM-65 Maverick, the Hunting BL755 (see Plate 4.4) and the Hughes Wasp anti-armour mini missile.

Maverick is a single-warhead rocket-propelled missile which seeks its targets either by television guidance, by imaging infrared or by homing on to reflected laser light (see Fig. 4.6). In attacking with the IR Maverick, the aircraft pilot would locate a target with the aid of an enhanced or electrically exaggerated infrared image on a CRT. The pilot then locks the Maverick on to the target by superimposing cross-hairs on its image; when the missile is launched, its on-board computer continually monitors the target's movements and adjusts the flight path as necessary. Whether using optical, IR or laser missile, the pilot is free either to depart or immediately engage another target. Up to six Mavericks can be carried for use against hardened targets such as armour or reinforced static targets.

The alternative mode of attack is based on the traditional 'shotgun' principle of compensating for aiming errors by covering the target area with submunitions, the

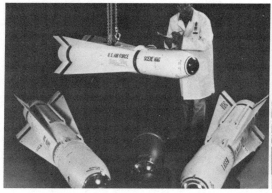

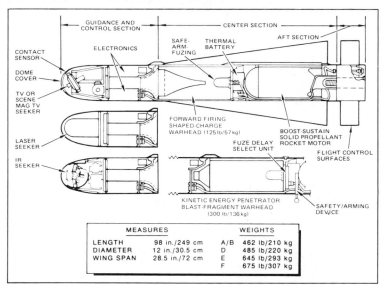

MEASURES		WEIGHTS	
LENGTH	98 in./249 cm	A/B	462 lb/210 kg
DIAMETER	12 in./30.5 cm	D	485 lb/220 kg
WING SPAN	28.5 in./72 cm	E	645 lb/293 kg
		F	675 lb/307 kg

FIG. 4.6. (a) A family of Mavericks. *From left:* The AGM-65F infrared-guided (IR) Maverick in development for the USN, the USA AGM-65D IR Maverick, and the USMC AGM-65E laser Maverick. Suspended for inspection is the AGM-65B TV Maverick. (b) A USMC ground crewman checks readiness of an AGM-65E laser Maverick missile prior to flight tests on a Marine AV-8B Harrier II. (c) Maverick missile arrangement.

latter-day shotgun pellets. In the British Hunting BL755, 147 shaped charge anti-armour bomblets are carried in each 'cluster' bomb (see Plate 4.5). After release from the aircraft, a preset timer fuse detonates the outer casing, releasing the submunitions. Each bomblet is slowed down by parachute, giving a higher impact angle, and explodes on impact, releasing a high-velocity jet and slug for armour-piercing. As a ballistic free-falling missile, the BL755 is immune to counter-measures, but does demand the aircraft to overfly the target or approach close to it. BL755 is particularly lethal when enemy ground forces are bunched, as in rendezvous, concentration area or choke point, and especially when softer skinned vehicles or troops are interspersed with armour.

Predictably, the microprocessor is bringing precision to the shotgun, volume and accuracy, in the terminally guided submunition. One such example is the Hughes Wasp anti-armour missile, designed for carriage by A-10A or faster jet strike/attack aircraft in either the interdiction or CAS role, and under development in 1986. Twelve Wasp missiles are packed in 2,000-lb pods carried as external stores. They can be fired in whole pod salvoes of twelve or singly, at low or high level. Initially on a ballistic trajectory, the missile is guided towards its target by a solid state millimetric wave transmitter and receiver. Cruise speed, derived from two sustainer rocket motors, and altitude are controlled by computers within the missile itself to produce the best ground coverage. Target is selectively preprogrammed, instructions and each missile can be independently targetted: all by self-contained digital processor. Finally, the target is attached by shaped charge warhead. In the immediate future, therefore, it is probable that one aircraft will be able to engage up to forty-eight tanks, without entering line-of-sight range, and with a high kill probability on each, despite the uncertainties associated with moving targets. Moreover, as will be explained in a later chapter, millimetric shortwave radar is difficult to jam.

AIR DEFENCE SUPPRESSION

Whether intervening directly in the land battle, or seeking to interdict reinforcement and resupply, the attacking aircraft will face an extensive range of hostile air defences, as explained in Chapter 3 above. Both attack and defence is likely to be shrouded in electronic warfare, as explained in Chapter 5, but the attacker can complement such protection by carrying special-type defence suppression weapons. Clearly, any precision weapon can be used against surface-to-air defence installations, but a family of anti-radiation missiles has evolved which uses for guidance the emissions from hostile radar targets. As explained in Chapters 2 and 3, surveillance, early warning, acquisition and guidance systems for many air defence weapons are dependent in some way on radar. The destruction of the radar component can therefore render the fire system useless or at best severely degraded. Since World War II, the principle has been well understood and weapon development proceeded apace.

Three separate missiles, developed over a period of twenty-five years, illustrate vividly the impact of microprocessor and solid state technology on a most complex interface of traditional weaponry and electronics. The first is the US AGM-45 Shrike anti-radiation missile (ARM), which entered service in 1964 and has been fired in combat in South-East Asia, the Middle East and the South Atlantic. Shrike has been modified many times in that period, but its basic character and mode of operation have remained relatively constant. The missile is driven by a solid fuel motor at speeds in excess of Mach 2 over ranges of 12–40 km depending on the variant, delivering a 66-kg blast/fragmentation warhead detonated by a proximity fuse. The missile is steered in flight by four gas-driven fins which respond to the direction of the hostile signal. It is, however, a preprogrammed missile useable against transmitters only on one frequency. It cannot, therefore, be carried for general purpose use or against unexpected frequencies, nor could it respond to modern frequency-hopping radars, and finally

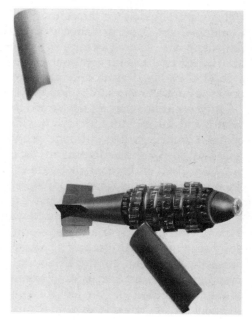

PLATE 4.5.

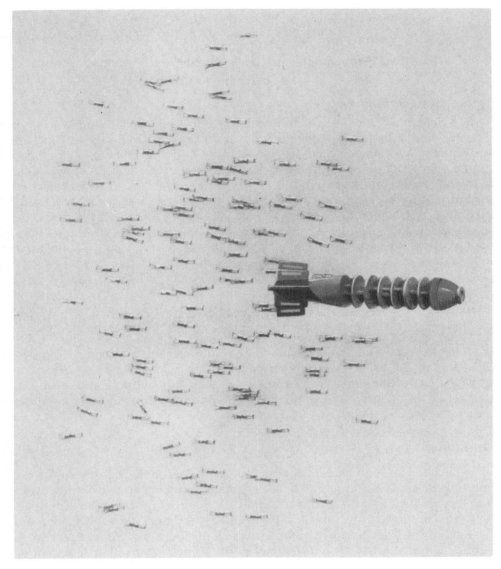

PLATE 4.5. Sequence showing delivery of BL755 weapons.

it has no memory. Nonetheless, it has been fired in anger in many areas, but with uncertain rates of success.

As a result, a second generation of AR missiles was produced, typified by the US AGM-78 ARM which became operational during the 1970s. AGM-78 had a higher speed over longer distances, but more importantly could respond to more than one frequency, had memory circuits and could make indirect approaches to the hostile emitter. It is possible that a variant of AGM-78 was used by the Israelis in their annihilation of the Syrian SAM installations in the Bekaa Valley in 1982.

By 1986, solid state technology was beginning to have a profound effect on ARM development in Europe and North America. The weapon chosen to equip the RAF Tornado GR Mk1 was the British Aerospace Air Launched Anti-Radar Missile (ALARM). Its operational principle, homing on to hostile radar emission, resembles that of Shrike, but in every other respect it is a product of the microprocessor revolution. ALARM is designed to be largely independent of systems on board the parent aircraft. It can be preprogrammed before take-off with a complete 'library' of possible hostile signals, from which a number can be selected and afforded an order of priority. The radar imprints will include surveillance, acquisition and missile homing and the combination of two programmes will allow ALARM to be fired without the aircraft's own radar warning receiver being alerted. It is widely assumed that ALARM not only has memory circuits, but is equipped to overcome the range of anti-ARM measures described in Chapter 5. There is, however, still more to ALARM. Because the missile itself has a database, its programmes can be updated by the aircrew during flight, considerably enhancing mission flexibility. Moreover, it can be launched in one of three modes. One, the traditional, is the direct attack on the radar emitter. The second, which is in fact the missile's primary mode, is indirect, during which the missile climbs automatically to 40,000 ft, deploys a parachute, loiters and searches for emitters. On acquisition and lock-on it dives unpowered onto its target. The third mode is a combination of the first two, whereby if the missile fails to acquire a relevant target in a direct attack it will revert to indirect, loiter search. Despite the complex circuiting required to sustain such electrical and mechanical manoeuvres, ALARM is considerably smaller than its predecessors, with corresponding reductions in weight. The net result is a quantum increase in self-protection for offensive support aircraft but reduced costs in terms of limitations on air-to-surface primary weapon load.

THE TOTAL IMPACT

In the late 1980s, a single artillery shell could deliver some 30 lb of explosive, a multiple rocket launcher several hundred pounds in a concentrated barrage over a few seconds. One aircraft could place up to 20,000 lb of high explosive in one pass over a single target array. Conversely, it could destroy single targets with pinpoint accuracy by missiles launched beyond the range of tactical surface-to-air defences. Increasingly, the attacks could be mounted by day and night and in deteriorating weather. Different target acquisition and missile homing sensors were being integrated into one fire control system. Tendencies in air defence were being matched in the aircraft designed to challenge them. Data management systems were emerging in which subsystems increasingly come to function automatically, thereby reducing aircrew work load in a similar manner to that of the fighter pilot. Overall, it was no coincidence that offensive air support was an integral part of NATO and Warsaw Pact war-fighting strategies.

5

Surveillance, Reconnaissance and Electronic Warfare

In the history of warfare, control of 'the high ground' has always been of vital interest to a commander. In addition to the natural advantages its possession conferred on the defence, it allowed him to see much farther. The farther he could see, the more warning he would have of an enemy attack and the greater his knowledge of enemy dispositions. The tradition of using the third dimension to gain even more height began when tethered balloons were used in the French revolutionary wars. The first military use of aeroplanes was for artillery spotting and reconnaissance in World War I. By the late 1980s, the functions had become many times more complex, demanding specialist systems which, like their offensive and defensive counterparts, drew increasingly heavily on the microprocessor and associated technologies. Moreover, air power signified the use of the air by man not necessarily with man, and the unmanned aircraft was already making a major uncontroversial contribution to tactical reconnaissance. Above them all, in the highest ground of space, unmanned satellites provided the major powers with a wealth of intelligence data which probably, on occasions, overstretched the capacity of the organisations on the ground to interpret and apply it. So much so, that one of the most dramatic examples of the fusion of technology and air power, the Lockheed SR-71 Blackbird reconnaissance aircraft, was more than once threatened with withdrawal from USAF service. A brief survey of the technological capabilities and limitations of reconnaissance satellites serve to explain why, in the event, the SR-71 continued to evolve and indeed has been joined by many other manned and unmanned aircraft within the earth's atmosphere.

SATELLITE RECONNAISSANCE

Reconnaissance satellites have been used by the United States and the Soviet Union since the early 1960s. The United States TIROS I spacecraft maintained an orbit 450 miles high, carrying two cameras and relying on solar cell batteries for power to transmit magnetic taped photo images back to earth. Both wide angle and narrow focus shots were taken, ranging from an area 135 miles long and 800 miles wide to specific shots of runways and missiles. By the mid-1970s, the resolution of such photography was good enough to identify cracks in runway or hardstanding surfaces. Contemporary quality was dramatically illustrated in 1984 when photographs of the newest Russian aircraft carrier were leaked to *Jane's Defence Weekly* which revealed the minutest details of the new vessel's superstructure in the Black Sea dry dock. It is believed that film images are both transmitted back to earth and ejected in heat-resistant cassettes from the satellites on command. The cassettes are controlled by small rockets for direction and re-entry to the earth's atmosphere, and emit radio signals to facilitate recovery. Other reconnaissance images are gathered by infrared and radar techniques with their information passed by data link to ground stations. The Soviet Union is reported to have at least two real-time ocean surveillance satellites in low earth orbit, reflecting her greatly increased interest in world-wide maritime affairs in the last three decades.

Increasingly, however, more than just imagery intelligence is demanded of reconnaissance. Eavesdropping on hostile radio emissions has since World War II acquired a considerable new dimension. Of continued significance is the intelligence to be gleaned from communications: between units at or under the sea, on land and in the air. Early warning of military activity can still be acquired from increased signals traffic while awareness of hostile command and control procedures has an obvious value. Now, however, many offensive and defensive weapon systems depend for their effectiveness on associated radar transmissions. If such transmissions can be detected, located and identified, they can increasingly be neutralised. Signals intelligence (SIGINT), also euphemistically referred to as Electronic Support Measures (ESM), is therefore an essential element in reconnaissance not just for traditional, conventional warfare, but for electronic combat also. Electronic counter-measures (ECM) and counter-counter-measures (ECCM) depend for their effectiveness on the provision of timely SIGINT.

The contribution of satellites to both imaging and electronic reconnaissance, although considerable, has several obvious limitations. While a satellite in a low north–south orbit can track within sight of most of the earth every day, its very predictability is making it more and more vulnerable to anti-satellite weaponry. The Soviety Union is reported to have successfully tested anti-satellite explosive devices and both superpowers have experimented with laser and particle beam energy weapons. Moreover, all satellites relying on real-time data link transmissions back to earth are dependent on well-known static receiver stations. Even if the satellite were to remain unimpeded, its progress can be easily tracked and image collection thwarted by camouflage or concealment, and SIGINT curtailed by communication discipline and suspension when the satellite is known to be within range. Consequently, peacetime surveillance can be limited by predictable and visible presence, quite apart from the natural constraints of cloud cover on optical systems. In the event of hostilities, reconnaissance satellites are becoming increasingly vulnerable and hence less dependable to military commanders. No doubt the offensive–defensive pendulum of military history will in due course incline in space also, but all the major powers have concluded that reconnaissance by manned aircraft must be retained for the foreseeable future.

STRATEGIC RECONNAISSANCE

The most well-known, and certainly the most effective, manned reconnaissance aircraft is the Lockheed SR-71 'Blackbird' of the USAF (see Plate 5.1). The SR-71

PLATE 5.1. Stealth ahead of its time: the USAF SR-71 strategic reconnaissance aircraft, capable of speeds in excess of Mach 3 at heights over 80,000 feet.

was designed in the 1950s and first flew in 1963, entering operational service in 1966 at Beale AFB in California. In 1976 the world speed record of 2,193.17 mph and sustained height record of 85,029 ft were claimed, but subsequent reports generally ascribe a top speed for SR-71 considerably in excess of that, over Mach 3, and an operational ceiling much higher than the published record. The airframe is largely built of titanium alloys with a corrugated skin which expands and smooths out in the ambient temperatures of 500°C met during flight. From the outset, the airflow through the two J58 Pratt and Whitney engines has been computer controlled through varying positions on inlet spike, ejector flaps and nacelle-mounted by-pass and bleed doors, appropriate to the range of speeds required. The airframe is a blended delta wing and fuselage whose basic instability is countered by an eight-channel stability augmentation system controlled by the autopilot to dampen airframe oscillation. The flat profile reduces radar-reflecting cross-sections and the heavy metallic black paint enhances both radar absorption and visual camouflage.

Official USAF sources have accredited the SR-71 with the ability to survey 100,000 square miles of terrain every hour. The exact nature of the sensors remains highly classified, but frequent open press supposition and technical press descriptions of available, relevant equipment lead to the conclusion that SR-71 probably carries very-high-resolution sideways-looking radars, long-range oblique photography camera, infrared linescan to detect heat emissions, radio and radar signal receivers and signal-direction finders. Down links are likely to carry the reconnaissance data in real time back to earth and on longer sorties farther away from friendly territory it is possible that an 'up-link' to a satellite may also be used. The original sensor, navigation and communication suites in the aircraft have been progressively modernised over the last generation and in view of the sensitive routes which it can be called upon to fly, SR-71 is likely also to carry the most comprehensive of ECM equipment. The USAF's 9th Strategic Reconnaissance Wing, still based at Beale AFB, operates ten SR-71s at any one time, with a further similar number held in reserve. Two are usually deployed at RAF Mildenhall in England and a further two at Kadena AFB, Okinawa. The aircraft use a special low-volatility fuel and are therefore supported by two squadrons of modified KC-135Q stratotankers. Indeed, as so often happens with a sophisticated military system, the less glamorous link is vital and potentially very vulnerable, as the KC-135Qs must fly a prearranged racetrack pattern to rendezvous with the very thirsty SR-71s. Nevertheless, for regular surveillance in peacetime and for the provision of real-time wide area reconnaissance in crisis and conflict, the SR-71 combines all the flexibility of the manned aircraft with the sophistication of the most modern airframe, engine and avionics technology to an unparalleled degree.

TACTICAL RECONNAISSANCE

In many air power activities, distinction between 'strategic' and 'tactical' can be imprecise and occasionally misleading. Aircraft designed for a 'strategic' role, such as the B-52 or the SR-71, can be used tactically to affect the course of a combat in one theatre of operations. Conversely, aircraft designed for relatively short-range activities, such as F-15s, could by the assistance of in-flight refuelling be ferried thousands of miles to influence the outcome of a regional crisis or small power conflict. Such influence clearly transcends the traditional association of 'tactical' with a limited area. Some aircraft originally designed for strategic reconnaissance have been adapted, with or without in-flight refuelling, to a tactical role. Extended endurance, designed originally to confer long range, can be equally valuable for lengthy reconnoitring, in a relatively small area. The best-known example is the USAF Lockheed TR-1, descended from the U-2 which first flew in 1955.

The U-2 was designed to fly at high subsonic speeds at heights in excess of 60,000 ft. Its basic operating principle was that the higher the altitude, the broader the swathe of vision beneath it. The spectacular destruction of Gary Powers' aircraft near Sverdlovsk in 1960 may or may not have curtailed overflights of the Soviet Union but the need for surveillance could not completely be met by satellite, nor by the sophisticated but very expensive SR-71. The increasing requirements of the United States for aerial surveillance, in South-East Asia and elsewhere, plus the increasing availability of more efficient and varied sensors, led in 1966 to the production of the U-2R, virtually a new aircraft, with increased wingspan, extended fuselage and increased endurance beyond 15 hours if necessary. Already these aircraft could transmit in real-time intercepted signals by data link either down to earth or back to an airborne relay. The U-2s could at the same time be tracked by a command guidance system at line-of-sight ranges up to

PLATE 5.2. USAF TR-1 tactical reconnaissance aircraft: a combination of traditional aerodynamic shape and ultra-modern sensors.

400 miles, by which sensors could be triggered without recourse to the U-2 pilot who simply had to fly the aircraft. Experiments were carried out with forward-looking infrared, forward-looking radar and X-band radar sensors.

In the mid 1970s, the USAF had studied proposals to acquire a tactical reconnaissance drone, built by Teledyne Ryan and Boeing, known as Compass Cope, for use primarily in the European theatre. In the event, the project was abandoned and a further variation of the U-2, the TR-1A, was accepted by the USAF instead. Entering service in 1981, the TR-1 is similar in size to the U-2R, with a 19-m fuselage and a 31-m wing span (see Plate 5.2). Powered by a single Pratt and Whitney J75-P-13B turbojet, TR-1 is reported to cruise at over 70,000 ft, with a range/endurance of more than 3,000 miles at high subsonic speeds, carrying a variety of interchangeable sensors in nose, fuselage mission bays and underwing pods. The sensors represent the forefront of Western reconnaissance technology.

The primary sensor is the Hughes Advanced Synthetic Aperture Radar System (ASARS) which relies on digital processing techniques to provide radar pictures almost as clear as high-resolution photographs of areas reportedly up to 80 miles abeam the aircraft's flight path. The imagery can either be imprinted on film within the aircraft for later processing on the ground or transmitted immediately to ground stations or relayed by a communications satellite. Some reports imply that the system computer can be preprogrammed with target array information so that the radar will switch automatically on cue from wide area to high-resolution narrow coverage. The transmission of relatively low-power signals over a spread of frequencies reduces vulnerability both to intercept and ECM, and relies on advanced processing techniques to receive and analyse the weak return signals. In addition, TR-1 carries traditional electro-optical cameras and signal-emission receivers.

In the immediate future, ASARS will be supplemented by two other systems which promise the capacity to fuse the reconnaissance and strike/attack communication functions. The Joint Surveillance and Target Attack Radar System (JSTARS) is believed in trials to have located moving targets as small as single tanks at ranges of over 100 miles. When fully operational, the Hughes Pave Mover radar will identify and process multiple targets at ranges up to 150 miles abeam the TR-1 while still emitting low power on multiple frequencies. The promise is of wide area, moving-target reconnaissance with a real-time data transmission and hence the potential of direct targetting of air and ground fire power. Meanwhile, the imaging data provided by JSTARS will be complemented by a new signal emitter locator known as Precision Location Strike System (PLSS). PLSS works on the principle of location by triangulation from three TR-1s working as a team, flying in friendly airspace but within passive range of enemy radar or radio emitters at distances of up to 500 miles. The position of each aircraft is fixed precisely by friendly ground-based direction-finders; the bearing and timing of the hostile emission to each aircraft is compared instantaneously and an exact fix on the transmitter obtained, even with the briefest of signal receptions. The location can then be passed to attacking aircraft which can subsequently be updated on moving targets directly from a TR-1. It may be assumed that the frequency preprogramming, priority and variable-response techniques associated with the development of anti-radiation weapons will be inherent in PLSS sensors also. Comparison with systems operational in the early 1980s serves to emphasise just how dramatic is the increase in effectiveness expected from JSTARS and PLSS.

Several air forces were still operating specialist reconnaissance aircraft equipped with cameras and occasionally, infrared and radar mapping image collectors. The importance of reconnaissance was fully understood, but the technology lagged well behind the operational requirements. In the event of conflict in Central Europe, the ground battle would be fought between defensively located NATO forces and numerically superior, fast-moving Warsaw Pact armoured divisions. It would therefore be essential for NATO commanders to identify the main thrust and reinforcement lines of the opposition and to locate the concentrations of armour and support vehicles necessary to mount and sustain the momentum sought in Soviet military doctrine. Reconnaissance aircrew still relied heavily on the 'mark one eyeball' and optical imagery to collect that vital information. Consequently, the recce aircraft had to negotiate heavy air defences, search for the enemy ground formations, acquire the intelligence, evade the defence again on return to friendly territory and then either break radio silence to report what had been found or wait until return to base to be debriefed and unload camera film. Thereafter the intelligence data had to be processed, assessed and communicated to the defending ground forces or the offensive air support aircraft. Delays of an hour or more were common, despite considerable expertise on the

PLATE 5.3. Display from the Thomson-CSF RAPHAEL side-looking radar (Thomson-CSF).

part of ground crew and aircrew. In one hour a Soviet armoured division could have moved 20 miles or more, other units considerably more, rendering an effective air strike based on the reconnaissance data highly problematical.

The first major steps towards real-time reconnaisance provision came with the deployment in Europe of the modified USAF RF-4C Phantom II: basically the same aircraft which had seen service ten years previously in South-East Asia, but, in the late 1970s, progressively equipped with first generation computerised systems. Side-looking airborne radar (as in Plate 5.3) increased the recce 'swathe' cut by the intruding aircraft, while a central computer controlled infrared line scan collection, laser target ranger/designator and a helmet-mounted sight which could provide the pilot with line-of-sight target designation and sensor cueing through the pilots' head movements. The central processor used Loran and inertial navigation inputs to pinpoint the exact position of the recce aircraft and hence the location of the potential ground targets. The information could be transmitted via data link either through a relay aircraft standing back from hostile airspace or, at shorter range, directly to a ground fusion centre for analysis and onward transmission. But even with that advance, the actual target marking and coordination with the attack aircraft had more in common with World War II than the later 1980s. The procedure was known as Strike Control and Reconnaissance (SCAR) and was basically pathfinding.

The recce aircraft would rendezvous with the offensive support units and lead them at low level towards the previously located target. The final phase of the attack would begin with the recce aircraft marking an attack reference point, at a predetermined distance from the target, by ejecting illumination cartridges which exploded with a brilliant flash and a puff of smoke. This was followed by the actual marking of the target by phosphorus rockets or photoflash flares, thereby giving the attack aircraft time to 'pop up', visually acquire the target and begin the missile or bombing run. The recce aircraft would then make a further pass within oblique camera range of the target to record the result of the attack and finally lead the attack aircraft back out of the threat area, all at 100 ft and at several hundred knots. Not surprisingly, the RF-4C crews became renowned for their flying skills. And, equally unsurprisingly, there were many sceptics who argued that no amount of aircrew skill and bravery could complete many such missions in the face of increasingly complex air defences, especially when the approach and marking techniques gave an equally good visual fix to the surface-to-air defences and any hostile air superiority fighters in the vicinity. Such scepticism encouraged the development of various kinds of unmanned reconnaissance vehicles to meet the ever-increasing requirements for tactical reconnaissance.

REMOTELY PILOTED VEHICLES

The description 'Remotely Piloted Vehicle', or RPV, is somewhat loosely applied to a wide range of unmanned aircraft which may be preprogrammed to fly a particular route or may indeed be 'remotely piloted'. Experiments with radio-controlled 'drones' have been carried out for over sixty years in the West; for example, in the 1930s when the RAF used a remotely piloted aircraft to demonstrate, primarily to the Royal Navy, the difficulty of locating and shooting down an aircraft attacking ships at sea. The Navy was quick to accept the lesson and experimented with its own 'drone', the Queen Bee derivative of the Tiger Moth biplane. During world war II over 14,000 target drones were delivered to the US Army and Navy. Later, interest in RPVs for intelligence gathering and reconnaissance was stimulated first by the vulnerability and associated political embarrassment of manned aircraft such as Powers' U-2 and later by the need of the United States for tactical reconnaissance in South-East Asia. A widely used target drone, the Teledyne Ryan BQM 34 Firebee, was quickly converted to a reconnaissance sensor carrier, reclassified as AQM-34, and by the war's end had flown over 3,000 missions over North and South Vietnam and neighbouring countries. Surprisingly, to the lay observer, the USAF later retired its RPVs and indeed suspended funding of a strategic RPV, the Boeing Compass Cope. The reasons behind the USAF decision, and a subsequent 'rethink' in the 1980s, lie very largely in the technology available in the two periods. By 1986, several countries were obviously in no doubt about the immediate value of the RPV as a reconnaissance vehicle, quite apart from its longer-term contribution to many other functions of air power.

AQM-34 in Vietnam

Contemporary accounts of RPV use in South-East Asia were circumscribed by security restrictions, but it was known that Firebees were used for photographic reconnaissance, electronics intelligence gathering, bomb damage assessment and even propaganda leaflet dropping. The saving of aircrew lives by their use was a particularly strong emotional advantage, as was their relatively low 4 per cent overall loss rate. They were controlled from specially adapted C-130 aircraft which were used to release the RPVs and then monitor their flights. It was subsequently disclosed by the National Security Agency, fifteen years later, that at one stage losses of Firebees had been as high as 75 per cent because of the ability of North Vietnamese signal intelligence specialists to intercept the programme messages. Moreover, the RPVs lacked operational flexibility, were very expensive and depended on the comparatively vulnerable C-130. It was believed that the C-

130 would be unsuited to the far more hostile enviroment of Europe, both in the air and on the ground. Consequently costs, operational limitations and vulnerability led the USAF to concentrate, as has been explained, on various manned reconnaissance aircraft alternatives.

RPVs in the Middle East

The continued value of the RPV was dramatically illustrated by its use by the Israeli Air Force in the widely publicised campaign in the Bekaa Valley in 1982. The Israeli aircraft industries built a small drone, the 'Scout', powered by a 22-hp engine, carrying a 38-kg payload with an endurance of 7 hours. The reconnaissance aircraft carried a gyro-stabilised high-resolution TV camera which from 3,000 ft could cover an area of 50 Km^2 with a zoom lens which could focus sharply on a 50×40 m rectangle. Other variants, as will be explained below, participated in EW activities. The Scout was launched by catapult and controlled from a ground station by a three-man team. The operator either monitored the RPV on its preprogrammed flight or actually 'flew' it by remote controlled joystick. When a potential target was reached, the Scout was manoeuvred to provide a TV picture of it. The information could then be relayed to whichever fire unit, in the air or on land, had to be brought to bear. Scout's very low radar, infrared and optical signature reduced its vulnerability to ground fire and, in the relatively small geographical combat area, with closely coordinated Israeli offensive forces and relative invulnerability to either electronic or physical counter-measures, the RPV system was an extremely important element in the location and ultimate destruction of Syrian air defences and ground forces in the short July 1982 conflict.

The Modern RPV

At the same time that Scout had demonstrated its operational capability, the rapid evolution of several complementary technologies, already described above in a manned aircraft context, hastened a reevaluation of the RPVs potential. Microminiaturised digital electronics reduce the size and weight of many fundamental components: gyros, activators, power supply units and integrated circuits. Meanwhile, it has become possible to install more than one kind of sensor in a single RPV. For example, the Lockheed Aquila RPV, under development for the US Army, will carry a high-resolution TV camera, a laser for target ranging and designation and multimode video tracker. The whole mission payload is to weigh no more than 27 kg. A further development envisages an alternative FLIR system instead of the current Westinghouse electro-optics for night and restricted daytime visibility operations. Unlike Scout and its contemporaries, Aquila is designed to operate in the dense EW environment of a European battlefield and carries two steerable EHF antennas to provide jam-resistant data link for real-time provision of information to more than one receiving unit. Modern airframe stealth technology has been incorporated in an ulta-low radar-reflecting cross-section and the whole aircraft has an endurance of 3 hours at a range of 50 km.

British RPV development has followed a similar pattern. The Flight Refuelling

PLATE 5.4. British Falconet RPV on launching ramp (Flight Refuelling Ltd).

ASAT/Falconet (see Plate 5.4) was originally produced as a target drone, but was adapted to carry reconnaissance payloads of up to 50 kg integrally. Powered by a microturbo jet, Falconet could reach 400 knots, depending on its mission and, while limited in endurance, had enhanced invulnerability because of its high speed. A later vehicle, the Phoenix, has been designed from the outset for use by the British Army to track hostile ground force movement and contribute to ground fire control (see Fig. 5.1). The specialised role has prompted the installation of a variable field of view thermal imager which can pierce cloud to relay detailed information for several hours from a 50-km range. The system can be developed further by the incorporation of semi-automated image analysis in the ground control units.

In sum, by the end of the 1980s the RPV will be making a significant contribution to the reconnaissance function of air power. Its advantages and disadvantages *vis-à-vis* the manned aircraft are likely to be much more clearly defined. It is interesting to note that airmen generally have been rather more cautious in forecasting its advantages, and providing funds for the deployment, than have army commanders. A cynical view attributes the attitude to a widely quoted, albeit perhaps apocryphal, observation by a USAF general that 'There ain't much promotion to be had with RPVs.' Ironically, some who do espouse the RPV case most passionately are among those who at the same time are most critical of what they believe to be exaggerated claims for similar technology when installed in manned aircraft.

Certainly, modern technology has thrown into sharper relief the manifest advantages of the RPV over the manned aircraft. These may be summarised as on page 88.

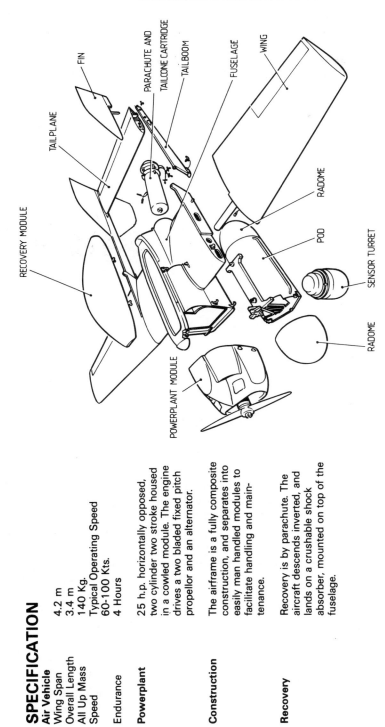

SPECIFICATION

Air Vehicle

Wing Span	4.2 m
Overall Length	3.4 m
All Up Mass	140 Kg.
Speed	Typical Operating Speed 60-100 Kts.
Endurance	4 Hours

Powerplant — 25 h.p. horizontally opposed, two cylinder two stroke housed in a cowled module. The engine drives a two bladed fixed pitch propellor and an alternator.

Construction — The airframe is a fully composite construction, and separates into easily man handled modules to facilitate handling and maintenance.

Recovery — Recovery is by parachute. The aircraft descends inverted, and lands on a crushable shock absorber, mounted on top of the fuselage.

FIG. 5.1. British 'Phoenix' surveillance and target acquisition RPV.

*Lower unit cost, and hence greater numerical availability per function.
*Reduced manpower requirements for operations.
*Reduced vulnerability in the air.
*Removal of aircrew losses to enemy action.
*Reduced training requirements.
*Reduced fuel requirements.
*No design limitations imposed by human physiology.
*Reduced maintenance requirements because of the smaller number of components.
*Most obvious beneficiary of microprocessor-related technology because of reductions in size, weight and power requirements.

Conversely, many limitations and disadvantages remain.

*Restricted payloads.
*Reliance on ground stations for data link inception.
*Vulnerability to jamming and other forms of transmission disruption.
*Relatively short range without intermediate airborne transmission link.
*Inflexibility in response to unexpected circumstances.
*Relatively role limited because of restricted payload.
*Increasing vulnerability in the air to modern high-speed, agile air-to-air and surface-to-air weapons coordinated with modern high-definition radar systems.

Nevertheless, it is probable that some of these disadvantages will be reduced still further by increased availability of systems using large-scale integrated, microminiaturised circuits. A 'robotic' air vehicle might be able to fly completely autonomous missions involving fully automatic position-keeping, target identification, reactive data link transmission triggered only by target identification, and some degree of preprogrammed in-flight decision-making in the event of a wide range of contingencies, followed by uncontrolled return to base. Under such circumstances vulnerability would be reduced still further, 'unexpected' events would be minimised, no in-flight control or communication system would be required, and associated ground stations would be very difficult to locate and destroy. Under such circumstances the RPV would become an extremely important complement to manned aircraft operations, although restrictions on payload and hence on role flexibility would remain. It is likely, therefore, that even the most futuristic RPV would remain primarily a reconnaissance or EW vehicle, but probably making a steadily increasing contribution to the acquisition of tactical reconnaissance.

ELECTRONIC RECONNAISSANCE

While traditional image and tangible data reconnaissance remains an essential requirement for military commanders, the acquisition of electronic data has progressively increased in importance as electronic warfare (EW) has evolved from the black art practised by a limited number of specialists in World War II to the pervasive, all-enshrouding character which it has now attained in modern war.

The contribution of electronic technology to air warfare has been alluded to in earlier chapters by reference to specific systems. The subject will be analysed in detail in a separate volume in this series, but it is convenient to summarise it by tracing its dependence on electronic intelligence. Frequencies outside the visual spectrum are now essential elements in many weapon and communication systems. Early warning, many air-to-air, air-to-ground, and ground-to-air weapon systems depend on electronic emissions for their effectiveness. Radio communications have obviously superseded visual signals for command and control in all but the tightest hand-to-hand combat. Increasingly however, electromagnetic emissions can be detected, distorted, jammed, and used as conduits for weapons directed at their source.

The tasks allocated to electronic reconnaissance can therefore be summarised as:

*To locate and identify the enemy's electronic order of battle: including surveillance and early warning radars, missile guidance frequencies and, if possible, his methods of countering friendly electromagnetic emissions. This collection of tasks is usually referred to as electronic intelligence gathering, or ELINT.

*To locate, identify and monitor enemy communications, known as communication intelligence gathering, or COMINT.

Upon reliable ELINT depends the effectiveness of all friendly counter-measures (ECM) and the preparation of counter-counter-measures (ECCM). Access to enemy communications in both peace and war can obviously facilitate military action at all levels from strategic to localised in-theatre engagements.

ELINT

Equipment designed to detect and identify air and ground targets, or to control surface-to-air, air-to-air and air-to-surface weapons, relies on sensors which translate physical variations into electrical signals. The most widely used sensor is still radar, which can in many circumstances overcome limitations on optical or electro-optical systems caused by poor atmospheric conditions or darkness. Pulsed electromagnetic waves are transmitted, reflected by an object and received by antennae usually in the immediate vicinity of the transmitter. Detection is achieved either by contrasts in reflectivity, for example an aircraft against the sky, or by Doppler effect derived from movement over a distance. The wavelengths are usually located in the centimetric or millimetric ranges. The task of ELINT is to acquire as much information as possible about the wavelengths and operational procedures used by hostile radars, and most of the larger air forces of the world have specialist aircraft for the task. For obvious reasons, however, ELINT (and COMINT) operations are tightly circumscribed by security considerations. Denial to an enemy of awareness about exactly how much you know about his electronic order of battle is an essential factor in establishing surprise and even supremacy in the critical opening phase of modern combat. It is apparent that the Israeli Air Force in 1982 was fully conversant with Syrian SA-6 and other weapon control radar frequencies, whereas in the opening days of the June 1973 war the IAF was taken badly by surprise because accurate ELINT on Egyptian SAM was obviously not available. The IAF used both manned aircraft and Scout RPVs equipped with ELINT sensors before and during the battles over the Bekaa Valley.

Nevertheless, despite security restrictions, many airborne ELINT systems have been described in the open press. The US Navy E-2C Hawkeye, in addition to providing early warning of air attack, has an automatic, computer-controlled passive detection system to detect radar emissions. Four receivers identify pulse width, pulse repetition interval, direction from transmitter, frequency and pulse amplitude. The ubiquitous computer classifies the signals, stores them in a data bank and relays them to communications data link for external analysis. An ELINT aircraft can be instantly recognised by the number of antennae located at different points on the fuselage. Hostile emissions thereby strike the antennae with minutely different modulations and intensity which are translated by computer into directional information. By comparing the signal with data already stored in the computer, its nature and probably the distance travelled from emitter can also be pinpointed.

The USAF RC-135, the SAF TU-95 Bear-D and the RAF Nimrod R Mk2 are among the best-known long-range ELINT and COMINT aircraft in contemporary operation. Bear-D is a regular visitor to the edge of British airspace, presumably monitoring RAF fighter control and intercept frequencies as well as compiling data on intercept procedures and associated communications. The use of large aircraft such as these permits the carriage not only of a wide range of sensors but also a number of specialist rear crew members. Tadiran Israel Electronic Industries, for example, is developing a system for use on aircraft such as Boeing Model 707 or Douglas DC-8 which is manned by two or three ELINT operators, ten to thirteen COMINT operators, a direction-finding analyser, a maintenance engineer and a mission commander, as well as the most sophisticated computer-ised equipment and secure real-time data communication links.

The collection of ELINT is not simply a matter of peace-time data accumula-tion. The function of the Precision Location Strike System (PLSS) carried by the TR-1 has already been explained. Smaller aircraft and helicopters such as the RU-21 and EH-60A Black Hawk are flown by the US Army to provide continuous ELINT during the ground battle by identifying and locating short-range enemy emitters, while the Soviet air forces deploy both Mi-8 Hip helicopter and An-12 CUB ELINT variants for similar purposes. ELINT equipment carried by two marks of the RF-4 Phantom illustrates the relationship between the timely acquisition of signals intelligence and the immediate application of counter-measures. The R-F4C II reconnaissance aircraft is equipped with, in addition to the imaging sensors already described, the Litton AN/ALQ-125 TEREC (Tactical Electronic Reconnaissance) system which is specifically designed to identify hostile missile guidance radars in the combat area. Signal data is automatically classified, the source located, and the information passed by data link to tactical commanders or to other aircraft in the area. Such a system provides for either threat avoidance or threat destruction. The F-4G Wild Weasel, on the other hand, carries a radar attack and warning system which can receive many signals at the same time, classifying them by type, range, bearing and threat priority. The information can be transmitted, but the primary role of the F-4G is to locate the threat and destroy it; moving from the task of ELINT to active, physical ECM.

ELECTRONIC COUNTER-MEASURES

The two-seater F-4G has been evolved to destroy hostile SAM and AAA sites using air-to-surface missiles designed to home on to the emissions of the missile and gun fire control radars. Whereas British aircraft will carry the ALARM anti-radar weapon described in Chapter 4, the Wild Weasels are likely to carry the US equivalent, HARM (High Speed Anti-Radiation Missile), or the general purpose AGM-65 Maverick, also described in Chapter 4. Physical destruction of the emitter conveys obvious advantages to the attacking aircraft, but it may not in fact be the most effective method of ECM.

Even assuming the guided weapon strikes the target—and even precision guided missiles miss occasionally—the warhead must detonate correctly and damage must be such that repairs are neither easy nor speedy. Moreover, the carriage of specialist anti-radiation weapons either calls for a dedicated aircraft or the reduction of other offensive weapon payloads to provide for the ARMS. Not surprisingly, therefore, there has been considerable investment in equipment and procedures designed to neutralise, rather than destroy, hostile electronic emissions. The MATRA system can be seen in Fig. 5.3. The design of any such ECM equipment depends upon accurate information about the hostile signals. The possession of a computerised data base permits the storage of vast amounts of detail about large numbers and types of emitters. Subsequently, the data has been used in the construction of preprogrammed ECM equipment such as 'chaff' tinfoil or equivalent, cut to correspond to specific wavelengths and present spurious returns on radar screens. With the advent of microminiaturisation, the data may be translated to the automatic systems which will identify, categorise and allocate priority to many different kids of emission. The extent and complexity of the required data base can be deduced from a brief summary of the range of electronic emission dependent threats faced by a modern aircraft in hostile airspace.

A Command to Line-of-Sight (CLOS) SA missile guidance system depends on both target and missile being tracked by a ground-based system which steers the missile by data link signal on to the target. Interference with the data link guidance is difficult, but the ground tracking radar is vulnerable to jamming. A semi-active homing missile, on the other hand, homes on to energy reflected from a ground-based or airborne transmitter. In either case, a radar signal receiver in the aircraft will register the illuminating signal and give notice of a missile launch. In the case of a semi-active missile system, the tracking radar can be jammed and the missile deceived into attacking a false target. The ECM equipment must therefore be able to identify incoming signals and counter them by generating responses of corresponding frequency, modulation and band width, with perhaps 30–40 sec warning time, depending on missile range.

The most modern 'fire and forget' guidance systems, such as those described in Chapter 2, present more complex ECM problems. The initial trajectory of such missiles depends on target information provided by the ground or airborne launcher platform. No early warning of 'lock on' is received by the target until the missile approaches within some 5 km, directed by its own integral transmitter and receiver. Such a missile is likely to use very-high-frequency radar band widths together with complex pulse modulations. The target will probably have only

SPECTRUM	SIGNAL	JAMMER	RESULT
HOPPING SIGNALS / JAMMER SIGNAL / 5 MHZ HOPPING WIDTH	Frequency hopping AM signal is over 5 MHz bandwidth at slow hopping rate.	A 2 MHz wide noise jammer in 5 MHz wide slow hopping pattern.	Communication is degraded (the "pops" heard before jamming is applied, due to use of AM modulation).
5 MHZ HOPPING WIDTH	Same frequency hopping signal as in recording number 8.	Noise jammer now has spread jamming noise over the full 5 MHz hopping bandwidth.	Communication is now completely denied.
225 MHz / 400 MHz	The hopping signal now utilizes the full 225–400 MHz bandwidth.	Noise jammer now has spread its noise over the full band.	Communication is not denied because noise per hopping channel is reduced.
HOPPING SIGNALS / JAMMER / 10 MHZ HOPPING BAND WIDTH	Frequency hopping signal is now over a 10 MHz bandwidth.	Follower-type jammer can easily keep up with slow hopping signal.	Communication is completely denied because frequency hop is jammed almost instantly.
225 MHz / 400 MHz	Same hopping signal as in recording number 11, but now over the full 225–400 MHz.	Follower-type jammer keeps up with slow frequency hopping signal over the full 225–400 MHz bandwidth.	Communication is severely distorted because most frequency hops are jammed almost instantly.

Frequency hopping radios now coming into use to provide more secure communications are susceptible to countermeasures of different types that can seriously degrade service or even deny it, depending on techniques used by jammer and communications systems, as shown by this Rockwell/Collins chart. Covert jamming does not deny service but increases user intelligibility problems.

PLATE 5.5. A Vulcan of 50 Squadron carrying Shrike anti-radiation missiles of the kind used against Argentine radars in the Falklands conflict (Flight Lieutenant Mike Jenvey, RAF).

10–15 sec to apply ECM, circumstances which dramatically illustrate the importance of an automatic response conferred by the use of very-high speed integrated circuits in the ECM system. Within that abbreviated timescale, the incoming radar signal must be detected, identified and reacted to. Whether the response is jamming, deception or, looking further ahead, an automatic input into the aircraft controls to initiate missile lock-breaking manoeuvres, there is little time for conscious, deliberate selection. In that analysis, the significance of a highly agile anti-aircraft missile is apparent.

Finally, the threat of the radar-laid anti-aircraft artillery (AAA) must be counted. AAA such as the Soviet ZSU 23-4 'Shilka' tracked guns are present in large numbers among Soviet ground forces, together with other radar-based guns of other calibres. Consequently, any NATO aircraft likely to be called upon to

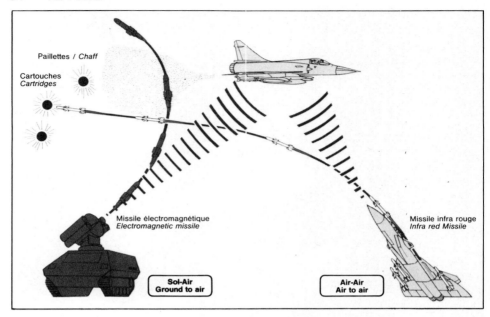

**Main characteristics of MATRA decoying
self-protection systems**

FIG. 5.3. MATRA aircraft self-protection system by self-dispensed decoys.

penetrate hostile airspace carrying ECM kit needs to be able to deal with AAA also. Tracking radar provides the guns with the target range and angle essential for initial accurate aiming. Moreover, at low and medium altitudes radar illumination is followed almost immediately by high-density fire, so that ECM response against the tracking radar must again be instantaneous. Denied tracking radar information, the kill probability of all but the most point-blank gunfire is considerably reduced.

ECM PLATFORMS

It is becoming increasingly feasible to equip a tactical aircraft such as F-16, F-15 (see Fig. 5.4) or Tornado with a comprehensive ECM suite, but both East and West have elected to deploy ECM specialists, particularly to disrupt hostile surveillance, early warning and communication emission. Such 'support' jamming demands very high power, reportedly at least a thousand times greater than that required for the self-protection equipment carried by a tactical fighter. Receivers need to be very sensitive to detect the threat radiation at maximum range, while associated signal processing must be able to cover a very wide band width at very high speed. As is explained in a later volume in this series on Electronic Warfare, the geometric relationships between support ECM aircraft, threat radar and its illuminated target are extremely important in the introduction of the jamming signals into the threat system. Not surprisingly, therefore, 'support' or 'stand-off' ECM aircraft have tended to be modified transports or larger combat aircraft. The Lockheed C-130H 'Compass Call' Hercules is used to jam hostile command, control and communication systems, presumably including those associated with airborne and ground-based surveillance and early warning radars. In the Bekaa Valley conflict of 1982, Syrian links between fighter controllers and their MiG-21 and 23 charges were allegedly jammed by 'stand-off' Israeli systems, probably installed in converted Boeing 707s, which had the power to flood the Syrian frequencies completely. In any future combat, such aircraft would become prime targets for ground- or air-launched anti-radiation missiles, inducing a very narrow electronic margin into the equation between miniaturised offence and defence.

Another USAF aircraft, the F-111, has emerged in a specialist ECM configuration: the EF-111A Raven (see Plate 5.6). Unlike the Compass Call C-130, Raven can either stand off from hostile air space or accompany friendly aircraft into it and is designed to counter not just communications, but a very wide range of threat-associated radars. It carries an ALQ-99E jamming system, an ALQ-137(v)4 Sanders self-protection jammer, a Dalmo Victor ALR-62(v)4 terminal threat warning system, an ALQ-23 radar counter-measures receiver system and a Lundy ALE-28 chaff/flare dispenser. Computers and miniaturisation permit the electronic warfare officer in the two-man crew to discharge a workload which ten years previously would have required several operators. The central data base is preprogrammed with known emitters and the automated ALQ-99E can detect, identify and assign jammers to neutralise enemy radars over a wide range of frequencies.

As a result, Raven can fulfil three separate but complementary roles. In the first, it could remain outside enemy air space and mask other friendly aircraft such as E-3 AWACS, TR-1 or reinforcement transports and tankers, by jamming long-range hostile surveillance radars. Second, EF-111As could move in close to the combat area and from low altitude jam the acquisition radars of enemy ground forces. Target tracking and missile guidance radars would be neutralised, effectively blinding the SAM and AAA batteries integrated with Warsaw Pact armoured divisions. The successful discharge of that function would considerably reduce the vulnerability of NATO's offensive support aircraft and increase proportionately their contribution to the air/land battle. In the third role, EF-111A would actually escort deeper-penetrating interdiction aircraft to their target areas, jamming continuously all the electronic elements of the air defence system, causing dislocation, delay, breakdown of coordination and a considerable reduction in missile kill probabilities.

Valuable as the stand-off jammer is, however, no major air force is prepared to depend on the availability of such systems every time a radar-associated threat is encountered by combat aircraft. In earlier generations, radar warning receivers alerted a pilot to the fact that he was being illuminated by a hostile radar. The signal was translated into an audible tone in his headset, with different frequencies represented by a spread of pitch and volume ranging from a low, comforting relaxed note representing friendly radar contact to a loud high-pitched continuous sound which signified enemy missile lock-on. In every case pilot response was

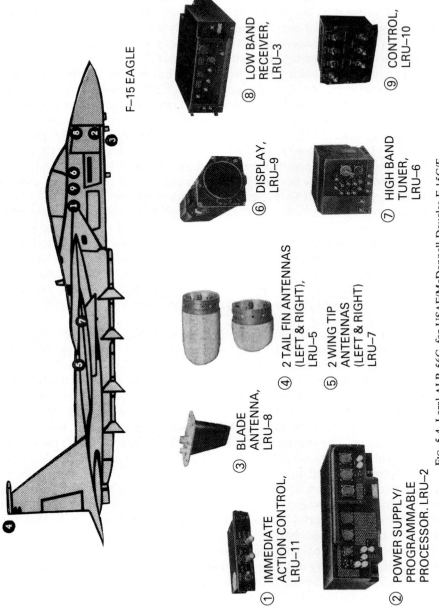

F–15 EAGLE

① IMMEDIATE ACTION CONTROL, LRU–11

② POWER SUPPLY/ PROGRAMMABLE PROCESSOR. LRU–2

③ BLADE ANTENNA, LRU–8

④ 2 TAIL FIN ANTENNAS (LEFT & RIGHT), LRU–5

⑤ 2 WING TIP ANTENNAS (LEFT & RIGHT) LRU–7

⑥ DISPLAY, LRU–9

⑦ HIGH BAND TUNER, LRU–6

⑧ LOW BAND RECEIVER, LRU–3

⑨ CONTROL, LRU–10

FIG. 5.4. Loral ALR-56C, for USAF/McDonnell Douglas F-15C/E.

PLATE 5.6. USAF EF-111 Raven ECM aircraft.

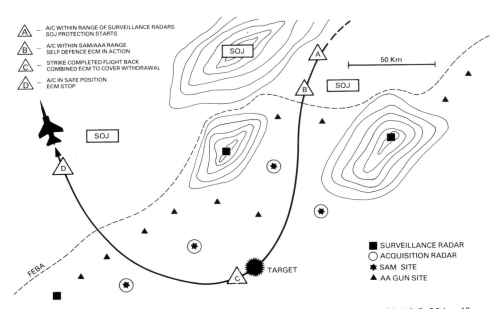

FIG. 5.5. Possible operational scenario for an aircraft carrying the SL/ALQ-234 self-protection pod, supported by stand-off jamming (SOJ).

deliberate and manual, ranging from violent manoeuvre, perhaps associated with chaff dispersal, to swift ejection seconds before missile impact. Now, however, integral or podded ECM equipment is carried by most combat aircraft.

One such system is the British Marconi Zeus (see Fig. 5.6), which incorporates a radar warning receiver with an automated jamming suite designed to counter ground-to-air missile tracking and guidance radars, air-to-air intercept and missile guidance radars, and AAA tracking and range radars. In addition to activating the Zeus jammer, the RWR can trigger the release of chaff, flares and decoys. Several other systems, such as the SL/ACQ-234 noise deception jammer developed by British, Italian and German companies, or the Westinghouse AN/ACQ-131 in use by USAF front-line aircraft designed on similar principles, are carried in pods on wing or centre-line hard points. It is likely that EW equipment, and especially ECM systems, will become an integral part of the aircraft rather than carried in external pods. Already the Zeus system weighs only some 110 lb and has the additional advantages of releasing a hard point for alternative external stores or reducing the conformal aerodynamic drag of the airframe. Thus, in both respects integral ECM systems enhance the performance of combat aircraft.

Not surprisingly, therefore, ECM systems are becoming design features in combat aircraft rather than expensive afterthoughts. In the interim, however, with many of the world's air forces either recently re-equipped or currently re-equipping with front-line aircraft, the immediate next ECM generation will be a further afterthought. Typical is the US AN/ALQ-165 Airborne Self Protection Jammer (ASPJ), under development for the USAF, USN and USMC. Designed as an internal system except for the USMC AV-8B, which will carry it in pod, ASPJ is fully automatic, dependent on computer control and comprises integrated receiver, processor and power management systems. It promises full coverage of continuous wave and pulse modes in all foreseeable threat frequencies. It is reported to be able to jam several threats simultaneously, using different transmission modes. As a system designed for a tactical combat aircraft, it is therefore unlikely to possess, or indeed to need, the power and range of the EF-111A.

Looking further ahead, it is probable that the next generation of US combat aircraft will carry ECM equipment under investigation by five competing US industrial teams. The concept, Integrated EW System, or INEWS, envisages operations over the whole electromagnetic spectrum: millimetre wave, infrared and laser as well as radar. ECM equipment may share computers and antennae with other avionics, and it is quite possible to predict fully coordinated, automated and integrated ECM-avionics-automatic flying controls by the mid-1990s.

ECCM

Indeed, the bounty of the very-high-speed integrated circuit seems limitless, until it is remembered that throughout military history, weapons and their associated tactics have swung backwards and forwards in an offensive/defensive pendulum. EW bred ECM: ECM has already begot ECCM. Raven carries a radar counter-measures receiver; Zeus claims control features which 'ensure that home-on jam weapons are not allowed enough time to acquire the aircraft'. Current technology: changing transmission frequencies faster than a jammer can follow; using so much power that a jammer cannot completely flood the signal; narrower main beams with reduced vulnerable 'spillage' into side lobes; refined information extraction from signals (for example, programming Doppler shift radars to exclude slow-moving response such as chaff); extraction of velocity information from conventional pulse radars; computer analysis of target dynamics to check data credibility: all are current or imminent elements in ECCM. No doubt by the time ASPJ and INEWS are in service, the counter-counter will be equally complex, equally automated and equally integrated. The electromagnetic spectrum will increasingly become a crucial battleground where the margin between victory and defeat may be narrow but very conclusive.

Undoubtedly influencing the outcome will be the acquisition of accurate and

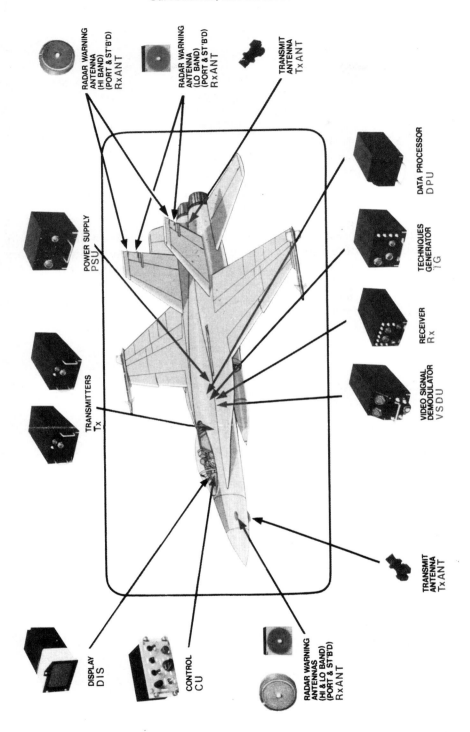

FIG. 5.6. Marconi ZEUS ECM system components.

PLATE 5.7. A French Air Force Mirage F1 fitted with a Thomson-CSF CAIMAN counter-measure pod (Thomson-CSF).

timely intelligence. Threat detection and identification have been the precursors of positive military action down the ages. In the later years of the twentieth century the sensors may change as the unmanned vehicle supplements the eye and the brain of the man. Traditional counter-measures of camouflage and deception will increasingly become the preserve of automated systems in ethereal as well as material combat. The nature of reconnaissance is changing rapidly as a result of the new technologies, but its contribution to successful warfare, particularly successful air warfare, is as indispensable as ever.

6

Maritime Operations

SEA POWER AND AIR POWER

Despite the exponential growth of commercial air transport in recent decades, the bulk of international trade is still carried by sea. Free passage on the high seas remains a jealously guarded international principle; as recently as March 1986 the United States was prepared to exchange fire with Libya over access to the waters of the Gulf of Sirte. But freedom of the seas can no longer be protected by naval vessels alone. The admirals' contention that the world is 75 percent covered by water remains true, but of greater significance is the fact that the whole of it is covered by air. Air power is as complementary to sea power as it is to any combat between ground forces. Indeed, in many respects technology has made contributions to maritime air power similar to those for operations over land. Consequently, several extremely important aspects of maritime air operations, and their associated technology, will not be dealt with at length in this chapter, not because of lack of appreciation, but because of their similarity with topics surveyed in Chapters 2, 3 and 4. Similarly, while there are still disagreements in several countries about the balance between the complementary roles of shore-based and sea-based aircraft, fixed wing and helicopters, this survey will cover aircraft, weapons and other systems which are deployed in the air on maritime operations regardless of the service which operates them and without reference to any role disagreements. The following sequence shows maritime technology in operation.

ANTI-SUBMARINE WARFARE (ASW)

The most obvious distinctive element in maritime air operations is anti-submarine warfare, usually fought in alliance with friendly surface vessels and submarines. Both East and West have invested heavily in anti-submarine technology, although for slightly different reasons. The NATO alliance and its international friends are spread across the world's continents. They depend for their economic strength on international trade, and for alliance strategy on the ability to reinforce Western Europe by air and sea from North America. Soviet submarines could pose a serious threat to both, either in limited or large-scale warfare. In addition, a proportion of the Soviet ballistic missile force is deployed on submarines and many of them must still venture well away from Soviet waters to move within striking range of North America. Finally, protection of Western warships against a submarine threat forms a major element in maritime strategy.

PLATE 6.1 Soviet Naval Air Force Maritime patrol aircraft, the IL-38 'May', 'marked' by a
Royal Navy Sea Harrier.

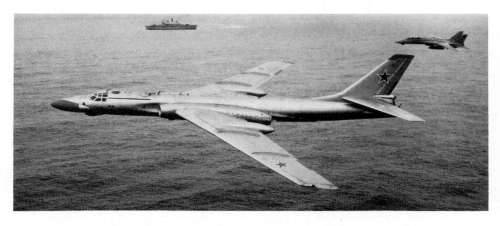

PLATE 6.2. Soviet Naval Air Force Badger reconaissance aircraft carrying traditional gun
turrets as well as modern radar and other sensors shadowed by USN F-14 Tomcat.

Meanwhile, the Soviet Union has laid great emphasis on anti-submarine
warfare to counter that proportion of the Western nuclear forces deployed under
the oceans and, with the rapid expansion of the Soviet's own surface fleet, to afford
protection against Western attack submarines. As in most comparisons of Eastern
and Western military technology, it may be broadly assumed that in areas

PLATE 6.3. Soviet Naval Air Force Tu-95 Bear-D reconnaissance aircraft, with several sensors clearly visible, shadowed by a RN Sea Harrier.

dependent on microminiaturisation, high-speed computing and very-high-integrated circuitry, the Soviet Union will tend to be some five years behind the United States and its allies. This survey, therefore, will concentrate on Western anti-submarine systems while the subsequent specialist volume will survey the whole field.

Airborne ASW is frequently shared between long-range land-based aircraft and shorter-range ship-borne helicopters. The fixed wing aircraft will have long unrefuelled endurance of eight hours or more, large carrying-capacity and a transit speed well in excess of any helicopter or surface vessel. Among the most effective ASW aircraft, if not the most, is the RAF Nimrod MR Mk 2 (seen in Plate 6.4). Nimrod is powered by four Rolls Royce Spey turbojets which confer a transit speed of 400 knots. On reaching the patrol or search area, two engines may be shut down and endurance extended still further. Nimrod's position is accurately sustained by an Elliot E3 inertial platform, a Decca 67M Doppler and a secondary Sperry GM7 Gyro Magnetic Compass system. In common with other long-range ASW aircraft, such as the Lockheed P-3 Orion, Nimrod uses a variety of sensors to locate the submerged submarine: sonobuoys, radar, magnetic anomaly detection (MAD), ESM, IR and ionisation detectors. On Nimrod, sensor information is coordinated and processed on board the aircraft by computer, and tactical decisions will be made by the crew members. On the Orion, on the other hand, analysis and decision-making is also computer assisted.

In World War II, airborne radar was the primary ASW sensor. Boats were usually diesel driven and consequently had to surface to recharge batteries as well as choosing to transit on the surface where possible to achieve higher speeds. Except for periscope observation, the modern nuclear-powered boat has no need to surface at all; even signal transmissions and reception can be carried out submerged. Diesel boats can recharge batteries with only the 'snorts' breaking the water. Consequently, while radar still has a significant role to play in long-range

PLATE 6.4. RAF Nimrod MR2 carrying AIM-9 air-to-air missiles.

detection, especially when submarines themselves are hunting at periscope depth, the primary modern ASW sensor tracks the submerged submarine in its own environment, by submersible sonobuoys.

Sonobuoys

The sonobuoy works on a simple principle. A hydrophone listens for the noise of a boat in the water and transmits any received signal to the aircraft above. The associated technology is complex. Any submarine moving through the water emits a sound, or sounds, from the rotating machinery of its engines. Those sounds, or signal frequencies, will be unique to a particular class of boat, and will comprise its acoustic signature. A single passive sonobuoy, i.e. one which is equipped only with a signal-directional receiver, will relay the signature back to the aircraft. If more than one buoy is used, the measurement of Doppler shift in the signals will give an estimate of both position and track. The buoy, retarded by a small parachute, is dropped into the water by the aircraft. A flotation pack is released which supports the antennae and power is provided by salt water-operated batteries.

Once a passive sonobuoy has given an approximate location, one or more active buoys may be dropped to give a precise fix. The active buoy transmits a short pulse of sound and registers the time taken for the 'echo' or reflection to return. With knowledge of the local water conditions, and the speed of sound through it, the distance and possibly the direction of the submarine from the buoy can be instantly calculated, thereby accelerating target localisation. The penalty, however, is that the target will be equally instantly warned that it has been detected. After a preprogrammed period of operation, a valve in the transducer below the buoy will open and the weight will drag the whole equipment to the bottom of the sea.

Typical of the modern passive sonobuoy is the British Dowty SSQ-904 Mini-Jezebel. An immediate impact on sonobuoy technology of the microprocessor has been the reduction of the basic buoy to one-third in size and half the weight of previous models without loss of performance. Operating details of Jezebel are highly classified for obvious reasons, but published data indicates that it can be launched at airspeeds of 60 to 325 knots at altitudes between 150 and 10,000 ft. It can remain active for preset periods of one, four or eight hours, has a frequency range of 5 Hz to 6 KHz, a selectable hydrophone depth of 18, 91 or 137 m and ninety-nine synthesised VHF channels. Other passive buoys, such as the US SSQ-53B DIFAR (Directional Frequency and Ranging), have greatly improved acoustic sensitivity in low-frequency ranges and electronic rather than preset RF channels, operating depth and operating period. Nimrod also carries the Dowty SSQ-963 Command Active Multi Beam Sonobuoy (CAMBS). Its hydrophone depth is adjustable by radio command over a thirty-one-channel telemetry link which uses common NATO frequencies. Directional data is obtained using a transmitter combined with a directional receiving hydrophone array from which bearing data can be extracted. It is particularly effective in acoustically noisy waters. CAMBS can be rapidly sunk by command signal should there be any threat to its security.

The submerged submarine can, however, seek to avoid detection by sonobuoys by 'hiding' among different thermal layers in the water which can distort and deflect the passage of sound waves. It is therefore essential for the hunter to know the temperatures in the search area. These are determined by a bathy-thermo-graph buoy dropped before the sonobuoys. The Dowty SSQ-937 miniaturised buoy has an operating depth of 427 m, descending at 1 m per sec, covering temperature ranges of $-2°C$ to $+35°C$. Data on temperature gradients is transmitted back to the aircraft for analysis and guidance on the depth setting for the sonobuoys. Thereafter, an alternative Vertical Line Array (VLAD) buoy may be used which has a set of omni-directional hydrophones attached at different depths with a DIFAR sensor at the phase centre of the array to provide target-bearing information.

Other Sensors

A helicopter, at the hover, can supplement sonar buoys by a dipping sonar, which is in effect a sonobuoy lowered into the water but which remains attached to the aircraft. It is therefore infintely reusable and can incorporate much more equipment than a disposable sonobuoy. A typical example is the Plessey Marine Type 195 used by Royal Navy Sea King helicopters. Type 195 can search through 360° and be manually controlled by the helicopter crew or be used automatically. A bathy-thermograph is incorporated, with the transducer allowing coordinated temperature analysis and operational depth settings. The operator has a choice of operating techniques dependent on sea conditions and the tactical situation and receiver range and bearing data. A new, highly classified, miniaturised dipping sonar particularly suited to the restricted space of the helicopter has now been produced known as Cormorant. Cormorant can be used in either active or passive mode and in conjunction with sonobuoys. The US Navy uses a long-range active dipping sonar, the AQS-13B, which is particularly useful in shallow water where extraneous noise is high. It has a specially shaped continuous wave pulse and has an optional Adaptive Process Sonar (APS) which provides precise digital data to enhance acquisition and definition of target bearing and range. As in the sonobuoys, technological advances have brought about both reduction in size and improved performance. The AN/AQS-18 is a smaller version of the AQS-13, with improved signal-processing techniques, higher transmitter power, longer range and, because of its compact nature, can be lowered and retrieved much more quickly. This last improvement is particularly important, because whereas sonobuoys can be distributed swiftly in a pattern over a predetermined area, the helicopter-borne dipping sonar must be lifted and lowered every time the helicopter wishes to change position. In this context it should be borne in mind that the helicopter will invariably be ship-borne and working in close tactical co-operation with the many ASW sensors carried by surface ships.

ASW Radar

As has already been noted, airborne radar does not now have a dominant position in ASW operations, but remains a most valuable component. If a submarine has been tracked by passive sonar, radar can pinpoint the position of a periscope or snort. As a result, several radars, for example the Thorn EMI Searchwater radar carried by the RAF Nimrod MR Mk 2, have been developed specifically to overcome problems inherent in an environment where the 'target' can be very small and probably obscured by clutter from the movement of the surrounding water. Frequency agility and a very short pulse width are used to reduce the extent of the reflected signal and increase the ability of the receiver to discriminate between target and clutter. Should a boat seek to take advantage of night or weather to surface, radars such as Searchwater will pick it up at considerable range. It can therefore be assumed that current generation submarines will carry a radar warning receiver on snorts and periscopes, and so EW becomes an integral part of ASW conflict as well as in the skies above.

Other ASW Sensors

Other sensors are used, and several are the subject of advanced experiments and development, but as yet are generally only of value to supplement or confirm the information received from sonar equipment or radar. Infrared surveillance can detect the heat generated by a submarine and in certain sea states a swiftly moving boat will leave a footprint of warmer water behind it which can last for several hours. The heat contact, however, will be virtually impossible to detect from above if the submarine is travelling at any depth, and even if near the water the IR sensor will need to be in its vicinity in the first place to detect it. Nevertheless, changes in ocean temperature caused by submarines remains a highly classified area of USN research.

Several aircraft in both East and West can be observed carrying a Magnetic Anomaly Detector (MAD) in a housing on the fuselage. But, as with IR, MAD is only effective when a submarine is travelling near the surface and the aircraft is at relatively low altitude. Even modern data processing is ineffective at ranges much in excess of 1 km.

Submarines may release chemical or radioactive traces which can be detected by sensitive equipment which can measure atmospheric content. While detection ranges can be much greater than IR or MAD, the data can also be much more ambiguous. A diesel submarine exhaust would, for example, be very difficult to identify off the coast of an industrial area and therefore equipment such as the RAF's Autolycus is very much a secondary sensor.

Finally, every submarine is faced with the problem of communicating with its own friendly forces. Trailing surface aerials for reception are difficult but not impossible to detect, but provided wavelengths are known to the hunters—hence again the importance of ELINT—transmission by a submarine becomes a very risky affair even in an age of compressed high-speed signals. ESM can therefore be a valuable additional aid to locating and classifying a submarine by a patrolling aircraft.

Sensor Analysis

Thus while Western admirals and air marshals concede that the oceans still remain largely opaque, and that the submarine is likely to retain a balance of invulnerability for some time to come, the risks of detection rapidly increase if a boat has to approach the vicinity of its target or if it has to come up to periscope depth to launch an attack. And, as technology is steadily enhancing the range and capability of sensors, so it is providing the means for the information which the sensors acquire to be presented, analysed and used as a basis for offensive response.

Just as a computer in a combat aircraft can store in its memory details of known enemy radar frequencies, so the ASW system can include a bank of known enemy submarine acoustic signatures. Now, the comparison of a single 'contact' with the data bank will be done automatically and instantaneously. Digital processors can equally quickly distinguish submarine signatures from extraneous noise by integrating data received at different times. The extraneous noise will continue to be random, but in many returns—ideally in all—there will be the constant signature of the boat. The days of the operator's headphones and glazed expression are long gone. Similarly, returns from several sonobuoys and other sensors can either be monitored independently or instantly coordinated by the central tactical computer and displayed on a CRT which is rapidly replacing the more traditional paper print-outs and makes available both current and recently acquired data.

Future developments are likely to increase the interaction between data reception and subsequent sensor control, for example in raising or lowering the depth of the transducer, changing any directional field and automatic tracking or, in an emergency, automatic sinking of a sonobuoy whose security is threatened. While the full electromagnetic spectrum will be exploited as far as possible, one major problem is likely to remain: salt water attenuates all but the very lowest of frequencies. The contribution of modern technology to ASW signal acquisition, enhancement and analysis will therefore become even more important as target data continues to increase in volume, well beyond the capacity of manual management. Miniaturisation will increasingly permit the carriage by helicopter of a quantity and variety of sensors hitherto possible only in the larger fixed-wing MR aeroplanes. Indeed ship-borne RPVs will be able to deploy clusters of small sonobuoys whose data could be relayed to ships, helicopters or land-based aircraft. Thereafter, depth charges, torpedoes and other anti-submarine weapons will be used with considerable enhancement of kill probabilities.

ANTI-SURFACE SHIPPING OPERATIONS

Offensive air operations against surface shipping share to a certain extent the broad technology used against targets on land. There are, however, several significant differences which have encouraged modified or specialised equipment. First is the problem of target location. Satellite and/or long-range aircraft maritime reconnaissance are essential for the location of hostile shipping. Whereas it may be confidently assumed, without close inspection, that an armoured concentration east of Berlin belongs to a Warsaw Pact country, no such assumption can be made about an unidentified naval vessel a thousand miles out in the North Atlantic, even though the whereabouts of friendly forces should be known to a patrolling aircraft. Second, when a naval surface vessel has been located and identified, there is little it can do to elude subsequent surveillance or attack. There are no terrain features to provide natural cover and no refuge in which to seek shelter. Conversely, there are not terrain features to shield the approach of low-flying aircraft either. Nevertheless, heavy sea states and bad weather can combine to present problems of target location and identification seldom found on land. Finally, fields of early warning and defensive fire are unrestricted. Such variations in reconnaissance, offensive and defensive conditions are reflected in air-to-surface weapon systems.

Aircraft

Characteristics of maritime reconnaissance and strike/attack aircraft are analysed in detail in the specialist volume in this series on Maritime Operations and Technology, but a broad survey is necessary to explain one of the influences upon weapons technology and re-emphasise the considerable importance of miniaturisation.

Of the four major maritime powers, the United States allocates the responsibility of air-to-surface shipping operations almost entirely to carrier-based aviation, the Soviet Union almost entirely to land-based, while Britain and France divide the responsibilities between the two. Carrier-borne aircraft are designed to operate from a compact base and the US Navy's two-seater Grumman A-6 Intruder is not only typical but is a useful example to illustrate the evolution of maritime air-to-surface systems into the current generation. As the A-6A, it entered USN service in February 1963. In 1985, it was announced that 300 new A-6F models would begin to embark on carriers in 1989. The original aircraft was designed to locate and attack targets at night and in all weathers, at low level, by conventional or nuclear bombs. By 1986, the ability of surface vessels to defend themselves against such close-range attacks at low level was becoming formidable as surface-to-air missiles and radar-laid guns of the effectiveness described in Chapters 2 and 3 above were deployed at sea. The A-6 had progressively been improved by the carriage of anti-radiation missiles, FLIR, laser range-finder and designator and a multimode radar using solidstate avionics. But it was still necessary for it to fly too close to a possible target to identify it, and to overfly it to drop bombs. The hazards of that kind of offensive operation were brought home to Argentinian attention during the A-4 Skyhawk bomb attacks on British shipping in the Falklands conflict, even when the A-4s could use the contours of the nearby land to mask their approach and departure. The new systems to be intalled in the A-6F are designed to allow it to detect, identify and attack either outside the range of hostile surface-to-air defences or with a modicum of protection against them. Synthetic aperture radar, of the kind used by land-based reconnaissance aircraft, will provide long-range ship identification. Instead of the traditional single picture received on a radar plan position indicator or other CRT, consecutive images are received, stored and then translated by digital data processor into a composite radar picture of the target area. A high-resolution image is created from which ship classification can be made manually or automatically from an intelligence data bank. Meanwhile, contemporary cockpit avionics akin to those carried by modern air-to-ground offensive aircraft and new uprated engines will complete the modernisation of an aircraft which will still outwardly resemble its predecessor some thirty years earlier.

At the other end of the maritime attack fixed-wing range is the Soviet Naval Air Force's (SNAF) land-based Tu-22M (Tu-26?) Backfire B. Whereas the A-6 has a 53-ft wing span, 54-ft fuselage, weighs 60,000 plus lb and can carry 18,000 lb war load over a nominal combat radius of 750 miles, Backfire B has a 113-ft wing span, 140-ft fuselage, 270,000 lb weight and is estimated to carry a nominal weapon load in excess of 26,000 lb over a combat radius of 1,500 miles. It may be assumed that Backfire does not carry the range of automated sensors available to Western strike/attack aircraft, but the much larger airframe permits the installation of powerful radars and no doubt comprehensive ECM and ECCM systems. The range is consistent with the Soviet perceived need to locate and attack US naval task forces before they come within missile launch range from their own strike/attack squadrons.

Two well-known aircraft in use with Western air forces were both designed for catapulted carrier-borne operations, but both are equally at home flying from land bases. The Hawker Siddeley Buccaneer was originally designed for Britain's Fleet Air Arm, but is now the RAF's primary anti-surface shipping weapon platform. Since its entry into RN service and subsequent product for the RAF, it has been progressively updated by avionic refits and stand-off weaponry electronic systems. Buccaneers can carry 16,000 lb of weapons in an integral bomb bay and on external pylons over a combat radius of well over 1,000 miles. The Super Etendard, which entered French naval service in 1978, has a shorter radius and smaller weapon-carrying capacity than the Buccaneer: some 500 miles carrying two drop tanks and one Exocet anti-surface shipping missile (ASM). The combination of aircraft and weapon has, however, proved effective in combat in both the Falklands and Gulf conflicts. For very short-range anti-shipping operations, helicopters are increasingly being deployed as in comparable activities over land and they are proving prime beneficiaries of the progressive miniaturisation of weapons and associated management systems. Finally, the Naval and Marine variants of the Hawker Siddeley Harrier (discussed in Chapter 4), Sea Harrier and AV-8B (see plate 6.5), as well as the SNAF Yak 36 Forger, are demonstrating the value of VSTOL aircraft for integral offensive and defensive support from smaller carriers than those required by the F-14s, F-18s, A-6 and A-7.

It is, however, indicative that to the casual observer AV-8B is not too different

PLATE 6.5. McDonnell Douglas AV-8B Harrier II advanced V/STOL fighter in USMC service (McDonnell Douglas).

in size and shape from Harrier GR Mk 1; Bear D/F from Bear A; and A-6F from A-6A. The USN's MacDonnell Douglas F/A-18, on the other hand, is manifestly the product of contemporary aerodynamics and engine technology. The F/A-18 Hornet is, like the USAF's F-16, a multi-role aircraft which can be fitted either as an air superiority fighter/interceptor or for the attack role. Hornet carries high-speed, high-capacity digital mission computers which can calculate various weapon trajectories which with several other stores-related sytems combine to make a bombing accuracy of 10–15 ft commonplace for experienced squadron pilots. The cockpit instruments resemble those of the F-16, using CRT displays for stores management, radar, engine monitor, data link, FLIR, laser spot trackers and navigation, while all primary flight instruments and weapon delivery data are projected onto the HUD. Nor should the similarity with the F-16 be surprising, because F/A-18 is designed for use by the US Navy or the US Marine Corps against shore-based, rather than maritime, targets, and to contest air superiority over a beachhead as well as over the fleet. With a speed of Mach 1.8, a combat radius of 200 miles and a ceiling of more than 50,000 ft, it is a formidable addition to a surface fleet's offensive power. But whether the aircraft is new, like Hornet, or progressively updated, as in the case of the other examples, the significant feature in maritime offensive operations is the dramatically increased kill probability of the current generation of weapons.

Air-to-Surface Weapons (AS)

The threat to surface shipping, including the heaviest converted battleship or aircraft carrier, is not just from AS weapons, from the missiles and heavy guns carried by hostile vessels of similar weight, or from the torpedo or missile from the submarine. It also lies in the handful of surface-to-surface missiles carried by little more than fast patrol boats. Such a threat was recognised by the US Sixth Fleet in the Mediterranean in the confrontation with Libya in March 1986 when Libyan patrol boats approaching within missile launch range where themselves promptly despatched. The ship-borne helicopter, carrying lightweight but deadly AS missiles, can extend both sight and reach of surface units without incurring the heavy costs of large-scale carrier operations. Conversely, supplementary use of AS helicopters can free high-speed longer-range aircraft to engage capital enemy units or targets on land.

In the Falklands conflict, Royal Navy Lynx helicopters launched eight Sea Skua missiles with 100 per cent success rates against Argentinian vessels. Sea Skua is typical of current generation helicopter-borne AS weapons: weighing only 145 kg, with a range in excess of 15 nautical miles, propelled by solid fuel motors, and guided by I-band semi-active radar against a target illuminated by the launch helicopter. It is a 'sea-skimming' missile controlled by a preset radio altimeter and in the Falklands proved its ability to destroy its target in very bad weather conditions. The critical system is the Ferranti Sea Spray radar installed in the helicopters which can detect and track small, fast navel targets (see Fig. 6.1). Its frequency, agility and monopulse beam complicates ECM, while digital signal processing facilitates multiple-target tracking, all functioning while the helicopter itself remains outside the range of most contemporary ship-borne surface-to-air defences. Not yet proven in action is a similar French combination: the Aerospatiale AS.15TT missile carried by the Dauphin II helicopter (see Plate 6.6). AS.15TT is a passive homer with a 30-kg warhead, guided by the illumination of the Agrion 15 radar over a range similar to that of Sea Skua.

The Norwegian Kongsberg AS Penguin contributes a different answer to the problem of attacking small surface targets, combining a sophisticated radar target acquisition by the parent aircraft with a passive IR seeker on the missile itself (see Figs. 6.2 and 6.3). This conveys a 'fire-and-forget' capability not possessed by either Sea Skua or AS.15TT and it carries a heavier warhead over a longer range. Penguin III is to be fitted to Norwegian Air Force F-16s in the anti-shipping role—in passing, reinforcing the versatility of modern combat aircraft—and promises launch altitudes of 500–30,000 ft at ranges over 50 km. Target data will be passed from the designator in the aircraft to the missile which, with a warhead weight of 120 kg, poses a threat to surface ships considerably larger than fast patrol boats. The IR seeker can distinguish ships against a land background more effectively than traditional radar-guided missiles; it can be decoyed by flares, but not jammed by ECM. Also in northern Europe, the multi-role Tornado has been adopted by the German Marine Air Force as its primary anti-shipping weapon platform (see Plate 6.7). It will be equipped with an uprated version of the MBB Kormoran, with an improved booster and sustainer rocket, digitalised guidance electronics, improved ECM resistance and a more powerful warhead. Combined with Tornado's own low-level performance, Kormoran Mk 2 will present a formidable threat to hostile shipping in the Baltic and North Seas.

Perhaps the best example of the marriage of two generations of technology is the carriage of the contemporary British Aerospace Sea Eagle missile by the well-proven RAF Buccaneer. The target is acquired by the aircraft's radar or from other sources and its bearing and range translated to the missile together with target selection data and appropriate ECM. The gas turbine engine gives the missile a range in excess of 100 km, guided by an inertial navigation system at a speed of approximately 0.85 Mach. Instead of a preset altitude, Sea Eagle flies at the lowest

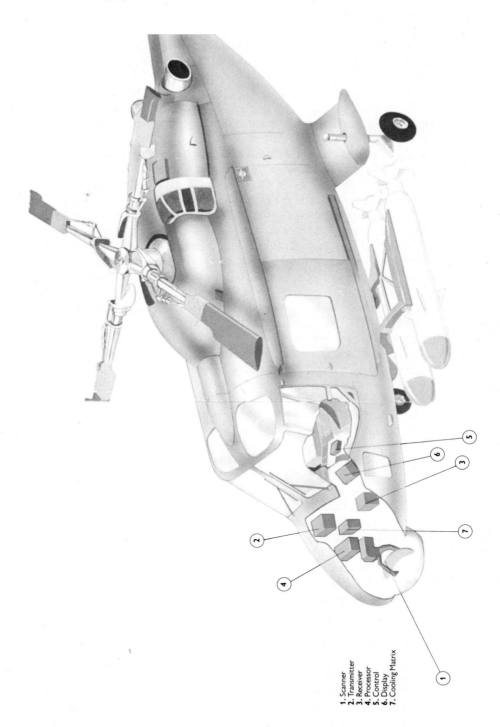

1. Scanner
2. Transmitter
3. Receiver
4. Processor
5. Control
6. Display
7. Cooling Matrix

FIG. 6.1. Seaspray Mk 3 installation in Westland Navy Lynx 3 with Sea Skua missiles.

PLATE 6.6. The Agrion 360 radar scanner and the compact derivative of the AS.15TT missile are clearly visible on this Dauphin II helicopter undergoing sea trials with the French Navy.

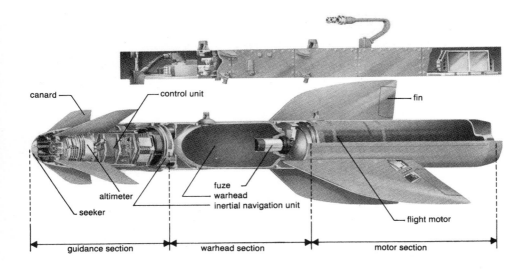

FIG. 6.2. Penguin Mk III anti-ship missile (for airborne deployment).

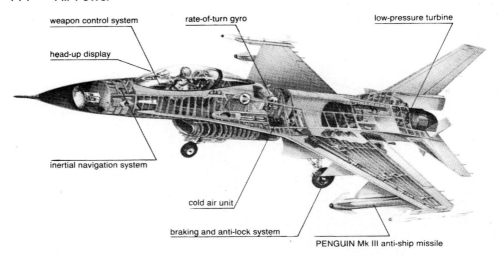

weapon control system   rate-of-turn gyro   low-pressure turbine

head-up display

inertial navigation system

cold air unit

braking and anti-lock system

PENGUIN Mk III anti-ship missile

FIG. 6.3. Penguin Mk III A5 missile on RNAF F-16.

PLATE 6.7. Panavia Tornado of the German Kreigsmarine at high speed over North Germany
(MBB).

height possible determined by a narrow beam width, low power and intermittent-transmitting radar altimeter. With a very small radar cross-section, a reduced IR signature because of the gas turbine propulsion and no active seeker transmission until the final stages of attack, Sea Eagle is not easy to detect, even by contemporary surface-to-air defences. A microprocessor in the missile controls the detection and selection of the desired target from any others in the vicinity. For saturation attacks, several missiles, each with a warhead of 230 kg, can be launched from more than one aircraft at one target or several without mutual interference. The onboard 16-bit Ferranti-F-100LB microprocessor probably provides a threat memory bank, triggering automatic responses to hostile radar acquisition and ECM.

Some indication of Sea Eagle's potential may be gauged from the success in combat of an earlier generation AS missile, the French AM39 Exocet (see Fig. 6.4)

Aerospatiale AM 39 EXOCET

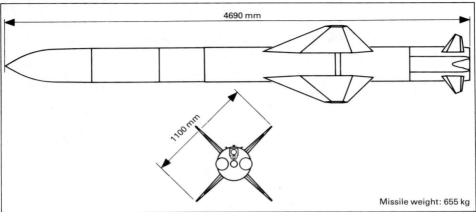

FIG. 6.4. Aerospatiale AM39 Exocet.

which sank HMS *Sheffield* and SS *Atlantic Conveyor* in the Falklands, two patrol boats in the Gulf and disabled several tankers in the same area. Exocet entered French service in 1979 and also depends on inertial navigation followed by active homing during the final stages of attack (see Plate 6.8). It has a maximum range of 70 km at high subsonic speeds and delivers a warhead of 160 kg. Like all the other stand-off missiles, Exocet does not require the launching aircraft to come within range of surface-to-air missiles, but unlike Sea Eagle (see Plate 6.9), the Argentinian Exocets at least were susceptible to ECM and increasingly vulnerable to short-range anti-missile missiles and high-intensity gun fire. Aerospatiale is therefore developing a 'Super Exocet', the ANS (*Anti-Navire Supersonique*), which promises a range of 200 km at speeds in excess of Mach 2, with very high manoeuvrability and advanced ECCM. The fact that defence analysts do not regard such promises with scepticism is a further reminder of the ability of designers to pack complex control systems, using programmable digital processors, fed by extensive memory circuits and powered by very-high-powered compact ram jet engines, into relatively small missile casings. The result will undoubtedly be to increase the vulnerability of surface vessels to air attack by reducing defensive response time and complicating problems of ECM. Conversely, the attackers will be able to stand farther off to launch the missiles and carry more of them with higher kill probabilities. In the past, longer stand-off ranges have presented problems of missile navigation, sometimes aggravated by marked changes in target speed and heading, or even by the emergence of a higher priority target after missile launch. Now, however, the combination of data link and digital processor permits the transmission of mid-course target update signals.

It would be foolhardy to assume that the balance of advantage at present tipping towards the AS aircraft and missile is likely to be permanent. Any technology which reduces weapon weight and volume must give a proportional advantage to a smaller platform, but the larger size of the surface vessel should allow the installation of an extensive range of defensive systems to reduce the threat of the

PLATE 6.8. Beyond the Super Etendard: a French Mirage 2000 combat aircraft carrying AM39 Exocet.

smaller aircraft. It is more likely that factors such as surprise, vigilance, human competence and indeed sheer weight of numbers will maintain their traditional influence over maritime operations.

AIR DEFENCE

It is readily apparent from the preceding paragraphs that air defence of surface units is a fundamental prerequisite for any attempt to achieve naval supremacy. Naval variants of short-range guns and SAMs, described in Chapters 2 and 3, are increasingly equipping moder surface ships. The E-2C Hawkeye and F-14 Tomcat (see Plate 6.10) were examined in the broader context of air defence in Chapter 2. Their significance in a maritime environment is, however, heavily underscored by the rapid expansion of stand-off AS weapon systems. A hostile aircraft must be intercepted before missile-release distance or the ship must be prepared to deal with the missile itself.

It is probable that the role of the EA-6B Prowler (see Plate 6.11) will be extended to contribute more to air defence. Hitherto, this specialist stand-off jammer, which discharges a maritime role similar to that of the EF-111 Raven over land, has concentrated on ECM support for carrier-borne attack aircraft, giving warning of, and directing ECM against, hostile surveillance, air-to-air and ground-to-air acquisition and guidance radars. There is an increasing need to jam the radars of

PLATE 6.9. Sea Eagle is a new generation air launched sea-skimming anti-ship missile under development by British Aerospace for use against warships with the most up to date air defence and electronic countermeasures capabilities. The missile is planned to enter service in the late 1980s, to arm Royal Air Force Buccaneers and Royal Navy Sea Harriers. It may later also be fitted to the Tornado GR1.

PLATE 6.10. The USN F-14 Tomcat in Atlantic weather.

surveillance and attack aircraft, as well as the active seekers of the missiles themselves. It was reported that EA-6B Prowlers were airborne throughout the confrontation between the US Sixth Fleet and Libya in March 1986. It is, however, unlikely that aerospace management over the open sea will become in the foreseeable future as complex as over land in the central region of Europe. At sea it will frequently be possible to assume that aircraft not under friendly control are hostile, and the fullest advantage taken of the range of missiles such as Pheonix. The maritime reconnaissance aircraft will become an increasingly

PLATE 6.11. The USN Grumman EA-6B Prowler. Under the port outer wing is an ALQ-99 jamming pod; at the top of the fin fairing is a pod containing high-frequency receiver antennae. The two flaps on the fin below the pod are for low-frequency antennae.

important target for the patrolling interceptor on the assumption that its destruction will considerably reduce the flow of target information reaching any attacking formations. Closer to land, however, the traditional vulnerability of surface vessels to air attack will continue to be most marked. Airspace management problems could begin to inhibit long-range air-to-air weapon release, the attack options available to land-based aircraft are increased, and ground contours may provide background cover for low-level missile attack. Conversely, surface forces operating in friendly waters should be able to draw upon the air cover available from land-based interceptors and missiles. Soviet attacks on Western surface vessels in the eastern Atlantic and North Sea, for example, should be complicated by the presence of extensive air defence forces in the European NATO countries.

COMMAND, CONTROL AND COMMUNICATIONS (C³)

Command, control and communications in maritime air operations present, with one major exception, similar problems to the exercise of C^3 in the air-to-air and air-to-ground environments. The exception is communication with submerged submarines. As has already been noted, sea water alleviates all but the lowest frequencies on the electromagnetic spectrum. The increased vulnerability of the submarine as it comes to or near the surface has also been explained. Even in traditional submarine operations, mounted against hostile surface shipping, boat captains were not inclined to surface before attacking, partly because of vulnerability to surface counters but increasingly because of the threat from patrolling ASW aircraft. They were operating in a state of conflict and their tactics, although frequently censured, were understandable. Moreover, the success or failure of a small number of boats—as opposed to the strategic impact of German

U-boats in the Atlantic and USN boats in the Pacific—was unlikely to affect the outcome of a major conflict. Forty-five years later, however, the submarine is a vital component in the deterrence posture of East and West. In the apocalyptic scenario of a nuclear conflict, and indeed in reassuring a potential opponent during uneasy peace, the perceived ability of a government to communicate with its nuclear missile-carrying submarines (SSBNs) is of paramount importance. But invulnerability of the SSBN before missile launch is vital to the credibility of the deterrence posture and thereby a difficult technological problem is presented. A ground radio station can communicate, using extra low frquencies (ELF), with a submerged boat down to some 180 m, but such a transmitter requires many thousands of feet of aerial and is therefore a prominent and vulnerable target for pre-emptive air or ballistic missile attacks. Moreover, even when the land station does transmit, ELF can only carry very restricted messages. The US Milstar satellite can also be used, but to receive the transmissions a SSBN would need to surface, thereby increasing its vulnerability and reducing its deterrent effect.

As a result, the US Navy has relied since 1963 on aircraft for communication relay with first its Polaris and then Trident SSBNs. A variant of the C-130 Hercules was fitted with a 200-kW VLF transmitter and a trailing antenna 10 km long. Over the last twenty years the ample cargo space and power sources of the Hercules have been used to provide progressive modifications to the system, known as Tacamo, which is popularly translated as 'Take Charge and Move Out'. Details of the system remain highly classified for obvious reasons. But however effective, the C-130 was restricted to a 7-hour patrol some 1,500 km offshore, which while allowing coverage of the Atlantic from both sides, left large areas of the Pacific without communication. Moreover, the effective life of the C-130 would probably expire by the early 1990s. As a result, the USN is developing a four-jet replacement based on the Boeing 707 airframe and drawing heavily on the digital communication systems of the E-3A AWACS, to be known as the E-6 (see Fig. 6.5).

E-6 is intended to have an operational range commensurate with that of the Trident force. It is designed to link 'backwards' with the Presidential E-4 and airborne command posts as well as downwards to the submerged Trident boats. Communication depth is reported to be 90 m. It must be able to relay emergency action messages, almost certainly in a dense EW environment and probably under nuclear threat. It is therefore hardened against the impact of high-density electromagnetic pulses of a kind associated with nuclear explosions and will carry integral ESM equipment. It will carry a primary 8,000-m antenna and a further 1,200 m wire as a dipole. To transmit, the aircraft flies a very tight orbit—so tight that aircrew seats are inclined to take account of the 50° bank—and the trailing wire drops to some 70° of vertical required for effective communication with submerged boats. In addition to the VLF transmitter, the E-6 carries VLF receiver, V/UHF, UKF, VHF and HF communication systems, together with a triple Litton laser inertial reference system, a Litton LTN-211 Omega and Smith's Industries Management Computer Systems for navigation; a Bendix APS-133 colour weather radar with tanker beacon homing and way-point display, two Honeywell radio altimeters, ILS, and a landing approach sensor. In addition to the

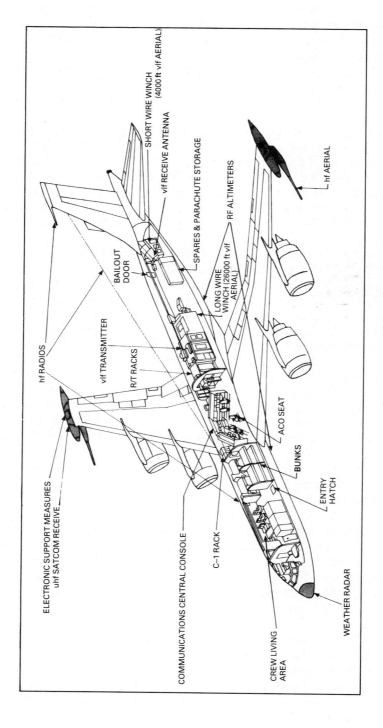

FIG. 6.5. The USN Boeing E-6 communications aircraft.

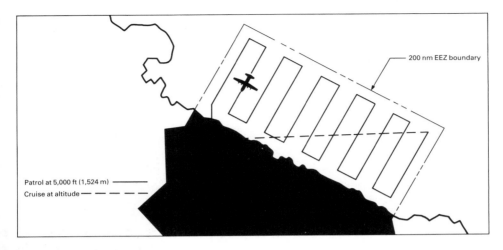

Fig. 6.6. A typical maritime patrol pattern by a C 130H-MP.

publicised data, it may be surmised that E-6 will also carry a very extensive range of self-protection electronics, probably data linked to the ESM sensors carried on each wing tip for automatic threat detection, analysis and possibly response. Fifteen aircraft are sought by the US Navy, for a total programme figure of $2,000 m. That figure is perhaps the most dramatic example of the possible costs of modern air power in deterrence; it is, however, certain that making provision for successful deterrence is cheaper than suffering the effects of its failure.

CIVIL AIR POWER

Far removed from the unthinkable and, hopefully, never-to-be-proven capabilities of the E-6, is the routine but nevertheless very important contribution of air power to the civil government. In the RAF, the sensors which equip Nimrod MR Mk 2 to hunt potentially hostile warships and submarines are also used in peacetime to monitor Britain's 200-mile Exclusive Economic Zone (EEZ) off the national coastlines. In the United States, the C-130, used by the Coast Guard for many years, is now being produced specifically for surveillance of EEZs. Fig. 6.6 shows an illustration of a typical maritime patrol pattern by a C-130H- MP. The capacious airframe can be fitted with any combination of modern surveillance sensors: search radar, side-looking radar, low-light television, infrared and ultraviolet scanner, FLIR, photographic and high-intensity search lights. Such equipment comprises a deterrent to more prosaic threats to national resources and illustrates the fact that maritime air power, like sea power itself, has a currency which is applicable in times of peace as well as in war.

7

Strategic Air Power

In World War II the expression 'strategic bomber offensive' was coined to denote the direct attacks on the German homeland by aircraft of the Royal Air Force and the US Army Air Corps. Whereas the term 'tactical' was generally applied to operations in support of either land or naval forces, 'strategic' thereby came to be associated with offensive air operations independent of any other service. With the creation of the USAF's 'Strategic Air Command' (SAC), the term became synonymous with long-range intercontinental attack directly against an enemy's home base and war-making potential. This chapter, however, will use the term 'strategic' in a much broader sense, and more specifically in contrast to 'tactical' operations. If 'tactical' refers to operations within a theatre of war, complementary to surface activity, 'strategic' air power is that which is applied beyond a single theatre or, occasionally, independently as a sole political instrument. The Israeli Air Force, for example, used air power independently against an Iraqi nuclear reactor, against hijackers at Entebbe and against a terrorist headquarters in Tunisia. In April 1986, in a controversial raid mounted simultaneously from bases in the United Kingdom and from carriers in the Mediterranean, the United States used air power independently of any surface forces to strike at several locations in Libya. To blur the issue still further, none of these examples were performed by aircraft of SAC, whose last 'strategic' activities were in the later stages of the war in Vietnam against targets in and near Hanoi. The Israelis used aircraft designed for relatively short-range 'tactical' operations, while the United States employed F-111 bombers which, although designated as tactical aircraft, in fact have a range which would in World War II have made possible attacks on the German homeland from UK bases. The conversion from tactical to strategic weapon systems is made by the use of in-flight refuelling, known appropriately as 'the force multiplier'. In the Tripoli and Benghazi raids of April 1986, the F-111s were refuelled on several occasions during their 5,800-mile round trip by KC-10 and KC-135 tankers. In exactly the same manner, transport aircraft designed for a strategic role—the C-5, the An-124, or the C-141—can be supplemented by aircraft of much shorter range but 'topped up' *en route*. Thereby, for example, the Hercules C-130 and even some helicopters can be called upon to mount long-range operations which more accurately can be designated 'strategic'.

STRATEGIC BOMBER AIRCRAFT AND WEAPONS

By the 1980s 'Strategic Bomber Aircraft' was a title more useful to denote a general idea than to identify accurately a class of warplanes and their equipment.

As has already been noted, a tactical fighter bomber could drop bombs on 'strategic' targets, while a strategic bomber was more likely to launch stand-off, long-range precisely guided weapons than old fashioned iron bombs. Nevertheless, the world's custom-built strategic bombers were readily recognisable. They included the USAF B-52 and B-1 and the SAF Tu-22M (Tu-26?) Backfire and Tu-95 Bear. All had the engine power, endurance and capacity to deliver heavy loads of conventional ordnance over very long distances—all of which could be extended even further by refuelling (although it should be noted that the SAF chose to remove in-flight refuelling equipment from Backfire). Alternatively, all could deliver nuclear weapons either by freefall trajectory or as warheads on stand-off, air–launched missiles.

B-52

Most well known in the West is the Boeing B-52 Stratofortress, whose prototype flew in 1952 and which entered USAF squadron service in June 1959. Since then several models of the B-52 have been produced, culminating in the B-52G and H, both of which were still in service in 1986. B-52H is powered by eight Pratt and Whitney TF-P.3 turbofans which generate more than 140,000 lb of thrust over an unrefuelled range in excess of 10,000 miles. It is 160 ft long, has a wingspan of 185 ft and can carry some 60,000 lb of conventional weapons at speeds up to 650 mph. Whereas the B-52H (see Plate 7.1) carries a 20-mm Gatling Cannon for self-defence, the B-52G relies on four machine guns. For both, however, the defensive armament would be very much a last resort after several other defensive systems had failed to ward off an enemy attack. Throughout its squadron service, the B-52 has progressively been re-equipped with modern defensive and offensive weapon technology. The large airframe capacity—the flight deck, for example, is more reminiscent of a civilian airliner in its scale than a traditional combat aircraft —has from the outset housed radar threat warning receivers and ECM equipment which have been replaced and increased as newer and more efficient systems have become available. The process was still continuing in 1986 with the announcement that the Litton ACQ-172 high-band jammer would be fitted. Some external indication of the range of electronic equipment carried by the later marks of B-52 can be gleaned from the antennae and aerials protruding from the fuselage.

Even in the freely informative Western press, exact details of the defensive equipment carried by B-52 are scanty, but airborne defensive systems available for installation in 1986 were well publicised. For example, the eight engines of the B-52 generate considerable IR energy and are vulnerable to IR missile attack; consequently, any jammer or decoy must radiate more intensely than the engines to afford protection. Most contemporary IR jammers use electrical energy to generate the radiation and if all eight engines were to be separately protected the drain on the aircraft's power source would be considerable. If, however, the direction of a threatening IR missile could be pinpointed, as well as its proximity as at present, energy could be saved and a more effective defensive system constructed. As it is probable that new generations of surface-to-air and air-to-air missiles will incorporate both radar and IR guidance systems, the spur to enhance IR counter-measures is sharp.

The long range of aircraft such as B-52 is at once a hazard and an advantage. A

PLATE 7.1. B-52H on unarmed training flight. Extensive range of electromagnetic sensors is indicated by antennae and aerials on lower fuselage, infrared sensor is in the 'pouch' below the nose.

hazard, because the aircraft can be allocated missions which will take it deep into hostile airspace with the consequent vulnerability to widespread air defences. On the other hand the possession of long range permits a variety of approaches to the final target area where defences may be more dense and opportunities for evasion more restricted. From the sources summarised in Chapter 5, the whereabouts of fixed surface-to-air defences *en route* are likely to be known and, if the hostile territory is very large, some high-threat areas can be avoided while others would be susceptible to ECM. The effectiveness of static defences in the region of the target themselves can be reduced, as in tactical air operations, by the launch of missiles beyond their radius of action. Since 1972, the major stand-off weapon carried by the later marks of B-52 is the Boeing Short Range Attack Missile (SRAM) (see Plate 7.2). SRAM is driven by two solid propellant rocket motors which give it a range of some 120 miles carrying either a nuclear or conventional warhead. It has an inertial guidance system which can be updated after launch, a terrain contour sensor and an onboard computer integrated with the B-52 weapon system computers. The missiles can be separately programmed before launch to attack different targets in a multiple launch at either high or low level. The aircraft does not need to acquire the target visually or by any other sensor, provided that target information relative to the aircraft's position is accurately known and after launch does not need to go nearer the target area and its associated defences. It is likely that Tu-22M(Tu-26) Backfire, although possessing only a fraction of the B-52 range, and not equipped with such advanced offensive or defensive technology, would also use stand-off weapons to attack strategic targets. Western journals

PLATE 7.2. Rotary rack of Boeing SRAM missiles in the bomb bay of a B-52.

identify two inertially guided air-to-surface weapons, both believed to be terminally homed by active radar seekers. One, the AS-4 Kitchen, may carry either a 200-kiloton nuclear warhead or 2,000 lb of explosive over a range in excess of 175 miles; the other, the AS-6 Kingfish, carries a similar warhead over a shorter distance, but at Mach 3 rather than Mach 2.

ALCM

But while SRAM, to a greater degree than either Kitchen or Kingfish, has profited from microprocessor development over the last generation, all have been overtaken by the stand-off weapon which is generally referred to as 'Cruise Missile', or in the USAF, Air Launched Cruise Missile (ALCM). A weapon which has entered SAF service, and is reported by the US Department of Defense to be similar in design to ALCM, is the AS-15, carried by the most recent addition to the Tu-95 fleet, Bear-H. The Boeing AGM-86B ALCM (see Plate 7.3) is in effect a very small, unmanned self-guided aircraft in which modern engine technology and miniaturised avionics have combined to create a weapon which presents considerable difficulties for air defences. ALCM is 20 ft 9 in in length, 4 ft in height and has a wing span of 12 ft. It weighs 3,000 lb and is powered by a Williams miniature turbofan engine which provides 600 lb of thrust and a subsonic range of over 1,500 miles. Designed originally to carry a nuclear warhead, ALCM relies on traditional inertial guidance, but with the novel addition of terrain contour matching (TERCOM). The missile scans the territory below it and compares the reality with the maps stored within its computer memory; then, if necessary, its course is automatically revised. It flies at very low level; it can be programmed for several route deviations and its very low electromagnetic signature presents problems both to look-down/shoot-down radars and surface-to-air detection and acquisition systems. The relatively narrow beam of the radar altimeter used by ALCM for ground scanning will give notice of the missile's presence, but in the immediate future will be difficult to jam. Moreover, at such low level the acquisition distances of all ground-based systems are reduced by basic laws of physics and degraded by terrain features. But as airborne radars increase in capability, as surveyed in Chapter 2, as weapon acquisition systems become much more responsive and as laser and other directed enemy systems begin to supplement more traditional air defences, even the relative invulnerability of the contemporary ALCM and AS-15 will decline.

Stealth Technology and Strategic Offence

As in so many aspects of modern air power, technology has been harnessed to very old military concepts to produce what has been greeted as a novelty and given the eye-catching title 'Stealth'. The word implies invisibility and is applied to measures designed to reduce the effectiveness of air defences; an earlier generation would have used the description 'camouflage'. An earlier generation would have used camouflage to reduce the effectiveness of one sensor: visual acquisition. Field grey uniforms, khaki uniforms, camouflage netting, mottled paint patterns: are all traditional methods of reducing the likelihood of a man, vehicle, ship or aircraft becoming visible to an enemy. Now, however, a camouflage is required which will reduce visibility across the entire electromagnetic spectrum. Stealth 'technology' is a misnomer, because it is in fact the amalgam of several different technologies. The clues to its development lie in earlier chapters which dealt with air defence, reconnaissance and electronic warfare. The cruise missile, by virtue of its small radar cross-section and low IR signature, embodies coincidentally some stealth technology; the successors to the

PLATE 7.3. USAF B-52 equipped with air-launched cruise missiles 'tops up' from a KC-135 tanker (Boeing Aircraft Company).

B-52 and Backfire have been designed from the outset with 'stealth' operations in mind.

The most visible elements in stealth technology are those designed to reduce the radar reflection of an aircraft. Right angles in an airframe structure and large flat surfaces are the major sources of that reflection. Obviously the aspect of the aircraft in relation to the radar emission source will vary, but by definition early warning radars are usually looking at the front in the critical periods of acquisition and identification. Whereas a flat surface or right-angled joint will reflect a radar signal directly, and hence most strongly, to the emitter/receiver, a curved surface and 'smoothed' corners will dissipate some of the reflected energy. If, in addition, the whole aircraft size, and particularly the frontal aspect, is reduced, the radar signature is reduced still further. The detection range of a target is proportional to the reduction in radar cross-section, but not equally, because the law of diminishing returns applies, as is explained fully in the ensuing volume in this series on Electronic Warfare. Consequently, other technologies are used.

Experiments in radar-absorbing materials have taken place since before World War II. Graphite-coated canvas, graphite-impregnated fibreglass, lightweight plastic or polyurethane foam, natural rubber with added carbon, neoprene, nitile and kevlar are among the materials used to reduce radar cross-sections. Combination of kevlar, fibreglass and graphite or metal would be lightweight and strong, but more traditional absorption techniques using elastomers have usually incurred weight and volume penalties. On surface ships and submarines, their use to reduce electromagnetic interference between different equipments and their environment, as well as to reduce radar cross-sections, has been cost effective, but in much smaller, highly-weight-sensitive aircraft choice of techniques is much more restricted. Radar-absorbent coating of traditional metals and modern composites is, for example, a technique already in use. It is highly probable that the SR-71 Blackbird's dull black finish is designed for more than optical camouflage. Such textures do, however, certainly impose a weight penalty on the aircraft and probably an increase in aerodynamic drag as well. The technique is therefore more suited to the airframe which is already designed to carry a heavy load and has the power to sustain it, as for example the USAF B-1 strategic bomber.

The third technique in the stealth evolution is the reduction of an aircraft's IR signature. As has already been noted, modern all-aspect IR missiles can detect aircraft surface temperatures as well as the more obvious radiation source of the engines. Nevertheless, a reduction of engine IR radiation increases the difficulties of the heat-seeking missile and recessed engines, air flow heat dissipation and aerodynamic shielding are already becoming more obvious in aircraft design.

Finally, an attacking aircraft must reduce the occasions it announces its presence by itself emitting electromagnetic impulses. Active target designators, continuous terrain-following radar and old fashioned radio emissions greatly facilitate the task of air defences. And moreover, in one critical aspect those defences, as has been summarised earlier, are likely to become even stronger. As the speed and processing power of computers continue to improve, the time between their target acquisition and weapon release will continue to shorten. As projectile defences come to be supplemented or replaced by directed enemy

PLATE 7.4. The contemporary strategic offensive aircraft: USAF B-1B on development trials.

weapons striking at the speed of light, detection, especially at low level or short range, will be even more swiftly followed by defensive riposte. On the other hand, the more effective the stealth technology, the smaller, in proportion, is the need for counter-measures, while passive IR, millimetric wave and non-continuous memory-backed radar emissions will reduce hostile opportunities for detection by signal reception. Consequently, the combined impact of the technology may be to encourage attacking aircraft to revert to higher altitudes, thereby providing more time to counter surface-to-air defences and easing several tactical and navigational problems associated with high-speed low-level operations.

B-1B

The airframe of the USAF B-1B strategic bomber (see Plate 7.4) has clearly been influenced by stealth technology. It is claimed to present a radar cross-section one-tenth of that of the B-52, and the curved fuselage lines and smoothed engine air intake angles are clearly visible. No official details of complementary stealth technologies as such have been released, but sufficient open source material exists to suggest that B-1B has a low radar reflectivity and is a formidable and flexible weapon system which will present difficult problems to opposing air defences.

With a four-man crew, B-1B is slightly smaller than the B-52, weighing 477,000 lb, 147 ft in length, 137 ft wing span and 34 ft high. It is powered by four General Electric F101-GE-102 turbofan engines of some 30,000 lb thrust, each of which gives a supersonic medium-to-high-altitude speed and a subsonic low-level penetration capability. Its range is classified, but the variable sweep wing to give aerodynamic efficiency across a broad speed range, plus its compatibility to refuel in flight with either the USAF's KC-10 or KC-135 fleet, indicates an intercontinental reach, or, conversely, extended patrol and diversion qualities. Unlike the B-52, it can operate in the manner of a much smaller aircraft from widely dispersed smaller airfields, thereby decreasing the vulnerability on the ground to pre-emptive attack. Physical protection against blast and thermal radiation has been built into the structure, while avionics, wiring and electrical equipment are protected against electromagnetic interference and lightning.

It is, however, its offensive and defensive avionics which mark it most significantly as a product of contemporary technology. The offensive system, designed by Boeing, is centred on six IBM digital computers which are netted with two microcomputers and additional processors in the Inertial Navigation System (INS) and Offensive Radar System (ORS). The ORS is itself based on the Westinghouse AN/ARG-164 radar developed for the F-16. All radar functions, including ground mapping, terrain following/avoidance, rendezvous and weather detection use one of two identical radar processing channels: the other being a reserve in case of the first's failure. The ORS transmits and receives through a new, phased array, Low Observable Antenna (LOA) which comprises over 1,000 electronically switched elements of which more than 10 per cent can fail without degradation of the system as a whole. The result is much improved radar mapping accuracy, several operating modes and reduced detection opportunity for hostile radars. The acknowledged inclusion of real-time signal processing within the system probably indicates the presence of a wide range of pulse frequency and modulation controls to enhance ECM. The complete offensive system has been designed from the outset to deliver a total payload of 125,000 lb which could include SRAM, conventional bombs, ALCMs and foreseeable generations of stand-off air-to-surface missiles. Information is presented to the offensive weapon systems operator by three computer-driven cathode ray tube displays which can show, on demand, either pictorial or written data. The Westinghouse radar picture is sharply detailed to allow pinpointing for accurate navigation and weapon delivery. The whole display suite is designed to facilitate rapid assimilation of target, threat and other data for appropriate response.

Defensive avionics have been produced by EATON/AIL and their details not surprisingly are less widely publicised. At the heart of the system is the ALQ-161 system (see Fig. 7.1) which is believed to be the most advanced and complex EW system ever installed in a combat aircraft. It was developed over a ten-year period and has been progressively modified to match the evolution of hostile radars. For example, a new high-frequency Band 8 jammer was added to jam Soviet J band tracking radars and further possible fits include new modulation techniques to counter monopulse radars and new antennae to improve emission direction-finding. ALQ-161 weighs almost 5,000 lb, but the voluminous capacity of B-1B has given the defensive system designers more scope than the limited space available to them in aircraft such as F-16 and F-18. It may be presumed that IR counter-measures, including jamming and decoy, are also incorporated in the defensive systems, but in early 1986 no authoritative unclassified information was available.

Details were available, however, of the advanced inertial navigational system installed in the new bomber. Later in this chapter other long-range navigational systems are examined, but the B-1B could well find itself operating a long way from friendly external navigation aids or indeed in an environment where such aids had either been destroyed or jammed. The Singer Kearfott inertial navigation system on the B-1 depends on an advanced accelerometer subsystem designed by Systron Donner of California. Eleven inertial sensor packages, combined accelerometer and gyro, monitor linear acceleration in both vertical and lateral planes and angular rates provide information on movement in all three axes to the aircraft's stability and control augmentation system. The systems have a sensitivity to a micro g of movement and provide exceptionally refined information on the aircraft's velocity and angular movement. Consequently, B-1B is likely automatically to maintain considerable navigational accuracy over very long ranges which, as in the case of all other strategic offensive aircraft, can be extended still further by air-to-air refuelling.

AIR-TO-AIR REFUELLING

The blurring of the traditional distinction between 'tactical' and 'strategic' air operations has already been noted. Foremost among the reasons for it has been the evolution of air-to-air refuelling (AAR) technology, first used on a large scale in the early postwar years to extend the range of USAF Strategic Air Command's bombers. Hitherto, the advantages of AAR had been recognised, but the technical problems of safely connecting a hose between two aircraft had not been overcome. The advent of jet aircraft with much higher speeds and, in the case of single seaters, no spare space for a refuelling hose operation, stimulated the invention in 1949 of a 'probe and drogue' method by the UK Flight Refuelling Ltd (see Fig. 7.2). This method is now the most widely employed outside the USAF, which relies on a flying boom in its KC-135s.

Refuelling Equipment

Throughout military history relatively simple steps in technology have often had far-reaching impact on operations; the probe and drogue AAR system is perhaps the best contemporary example in any environment otherwise inundated with mass computery and electronic extravagance. The tanker aircraft, which in every case so far have either been based on or adapted from well-proven airframes, carry a length of flexible hose wound on to a rotatable drum. One end of the hose is connected to the fuel reservoir tanks, the other has a reception coupling fitted with a conical para drogue resembling a large funnel. At the beginning of the AAR operation, the hose is extended gradually, restrained by an electric motor on the drum which will retract it on completion of fuel transfer. The receiving aircraft is fitted with a forward-facing, projecting rigid tube, known as the probe, at a point on the fuselage where it is clearly visible to the pilot. Then 'all' the receiving pilot

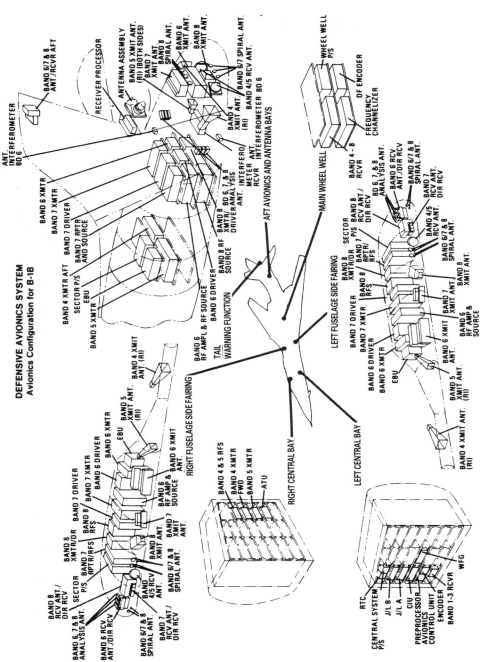

FIG. 7.1. ALQ-161 defensive avionics for USAF/Rockwell B-1B, built by Eaton's AIL Division, are a complex airborne electronic warfare system, consisting of more than 100 line replaceable units that collectively weigh about 5,000 lb. System cost is about $20 million.

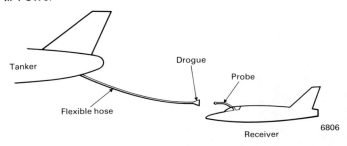

Fig. 7.2. Flight Refuelling Ltd probe and drogue method.

has to do is slot his probe into the para drogue. In practice, that calls for very precise flying to ensure that the probe enters the 30-in diameter cone at a speed high enough to enter and be gripped by the reception rollers at the apex of the drogue without being so high that the cone itself is knocked away before lock-in can take place. The torque on the hose is flexible enough to retract and absorb a first lock-in, but considerable flying skill and accuracy is required by the crews of both aircraft.

When the probe nozzle has made positive contact, the fuel is automatically released from the tanker and flows under pressure into the receiver at up to 5,000 lb/min (see Plate 7.5). The entire sequence is controlled visually by a panel of lights on the tanker fuselage on either side of the hose port in front of the receiving pilot. Red warns the receiver not to engage; amber indicates the hose is fully trailed and may be engaged; green indicates a positive engagement and fuel is flowing; while flashing red warns of an emergency and directs the receiver to break contact immediately and stand off from the tanker. By such a simple method radio silence can be maintained throughout and hostile ESM denied. At the completion of the AAR sequence the receiving aircraft simply reduces speed and the probe disengages.

RAF tankers carry, in addition to the central hose and drum unit, underwing pods which can refuel two other aircraft independently. The most recent aircraft, VC-10K and Tri-Star K1, will carry the MK 32.2800, an electronic, digitally controlled pod which requires no hydraulics and, in common with other more dramatic advances elsewhere, uses one-third of the components of its predecessors and provides fuel from the main tanks at a rate of some 2,800 lb/min. Of great significance to both engineers and planning staffs is the considerable increase in reliability and decrease in maintenance effort achieved by the new system.

The flying boom, on the other hand, as fitted to the USAF's KC-135 and KC-10 (see Plate 7.6) (which also carries hose and drogue), can only be used from the tanker's fuselage. It allows very high fuel transfer rates with the tactical benefits of minimum diversion time for the receiving combat aircraft and consequent acceleration in the number of aircraft able to refuel in a given period. It requires, however, a boom operator in the tanker who 'flies' the boom using remote controlled aerofoil surfaces, to make contact with inset receptacles on the receiver aircraft. A further limitation on the boom method is that the system cannot be 'strapped on' or inserted into a modified transport aircraft; it requires a dedicated tanker aircraft.

PLATE 7.5. RAF Victor K Mk 1 refuelling Hercules C-130 (Flight Refuelling Ltd).

PLATE 7.6. McDonnell Douglas KC-10 Extender tanker/cargo aircraft.

The most modern tanker in the USAF, the McDonnell Douglas KC-10A, is unusual in that it is equipped with both hose and drogue and flying boom equipment. No longer does the boom operator need to stretch out on his stomach in the tail of the aircraft to 'fly' the boom; instead, with the aid of the computer and information systems, he can sit at a console and operate remote controls. The possession of both kinds of fuel-transfer systems confers considerable flexibility on the KC-10A; able to refuel all USAF and Allied aircraft fitted for AAR in either mode. The manufacturers promise delivery of 200,000 lb of fuel to a receiver at a range of 2,200 statute miles. In the USAF attack on Libyan targets in April 1986, a large number of KC-10As were observed and reported in the British media to be in the United Kingdom. Without them, the F-111s could not have flow south round Gibraltar along the Mediterranean to Libya and return: the attack would not have been possible.

A further example in the AAR world of the impact of incremental rather than highly dramatic technology is the modernisation of the earlier USAF tanker, the KC-135, by installing new engines. The modified KC-135 'R's are driven by four CFM 56 high by-pass turbofan engines jointly developed by SNECMA of France and General Electric of the United States. At the same time, the aircraft have been structurally refurbished to extend flying life well past the year 2000. The KC-135R uses up to 27 per cent less fuel than the original KC-135A, which means in effect two can do the work of three. The modern engines consume far less fuel, leaving proportionately more to transfer over a much wider operational radius.

Even the most cursory summary of the operational benefits ensuing from AAR illustrates the exceptional significance and dramatic impact on air power of this rather mundane technological evolution:

- Offensive aircraft designed for tactical operations in theatre may be given a strategic capability.
- Fighter or reconnaissance patrols may be sustained for long periods without the need to return to base.
- Under threat of attack aircraft may take off from an airfield and be 'held' airborne for survival or delayed counter-attack.
- Tactical strike/attack aircraft may hold extended airborne readiness states to provide rapid reaction air support to ground operations.
- Long-range escort to strategic offensive aircraft can be provided by refuelled tactical fighters.
- Range extension to any aircraft increases the commander's options for variable routing to the target or any other destination, or simply for deeper incursions into hostile airspace.
- Range and mission endurance become limited only by aircrew fatigue and any need to rearm.
- Aircraft can take off either with lighter all-up-weight, permitting greater opportunities for dispersal and short field operations, or with a greater proportion of all-up-weight being consumed by weapons rather than fuel.
- Some airframes can be adapted for multi-role operations, for example, USAF KC-10A and RAF Tristar, both of which can be flown as tankers or cargo aircraft.

AIRLIFT

Air power is traditionally associated with the direct application of military force in or from the skies, but an important, if less fashionable, element in it is the use of air transport to give extended reach or resupply to surface forces. In the last two decades the advent of the wide-bodied cargo aircraft, equipped with modern turbofan engines and guided by contemporary and rapidly advancing navigational technology, supplemented if need be by AAR, has changed the face of air power at least as significantly as any other development.

Airlift was widely used in World War II, notably to drop paratroops as in Normandy or Crete, or for long-range supply of ground forces as in South-East Asia. Relatively short-range and limited cargo capacity compelled the use of large numbers of aircraft which were highly vulnerable to hostile air defences and consequently local air supremacy was usually a prerequisite for such operations. Indeed, even in the late 1980s, just as maritime transports required protection from hostile raiders, so the use of airlift on anything other than a clandestine scale would continue to require at least temporary and localised command of the airspace through which transport aircraft had to pass. In 1947, 347 aircraft from Britain, France and the United States mounted a peacetime airlift to Berlin for fifteen months which supplied the beleaguered city with $2\frac{1}{4}$ million tons of coal, food, raw materials and other goods. In 1973 the United States mounted a large-scale airlift to support Israel in the October War in some 570 C-5 and C-141 sorties; no resupply by sea reached Israel before cessation of hostilities. In 1986 the Berlin load could have been carried by seventeen Lockheed C-5A Galaxies, and no major power had any doubts about the potential of long-range airlift as an instrument of foreign policy.

Aircraft

The workhorse of the Western world's airlift fleets is still the Lockheed C-130 Hercules (see Plate 7.7). Nominally a 'tactical' transport, the C-130 has a range of over 2,000 miles carrying a payload of 45,000 lb. C-130, like most other specialist military cargo aircraft, is readily distinguishable from its civilian airline counterparts by its rear loading doors or hatch and ramp to permit ready loading and despatch of palletised cargo and vehicles as well as troops. Four Allison T56-A-15 prop-jet engines drive the aircraft at cruising speeds close to 400 mph. The aircraft is totally pressurised at 8,000 ft and has an undercarriage designed for landings on unmade strips of less than 1,000 yd. The transmission from 'tactical' to 'strategic' operations is best illustrated by the RAF's use of Hercules during the Falklands War, and subsequently to sustain the British garrison, when AAR flights of 13–25 hours were commonplace between Ascension Island and the South Atlantic and return. A second Lockheed transport, the C-141 Starlifter, carries a heavier payload, 90,880 lb, over a slightly longer range at 560 mph. C-141 can also deliver 155 paratroops, or 200 men with normal equipment, and was until the 1970s the most effective example of an aircraft designed for strategic transport operations. Since then however, it has been surpassed by two others, one American and the other Russian, which most dramatically illustrate the surge in and impact of wide-bodied military transport development.

The US C-5B Galaxy (see Plate 7.8) is the most recent variant of the Lockheed C-5 which first flew in 1968. Over a 3,300-mile unrefuelled range it can carry 291,000 lb of cargo at a cruising speed of 500 mph. At two extremes of fit it can lift

PLATE 7.7. USAF C-130 demonstrates its ability to take off from an unpaved strip.

either 340 troops or two M1 Abraham Tanks; or several combinations of troops, armoured personnel carriers and other vehicles. Unlike earlier military transports, C-5 can load via ramps at either end of the fuselage into a cargo hold almost three times as long, twice as wide and $1\frac{1}{2}$ times the height of that of the C-130. After fatigue problems in the wing designed for the C-5A, the C-5B wing is made largely from a new aluminium alloy which has increased resistance to fracture and corrosion. As a result, the maximum take-off weight of the aircraft has increased from 769,000 lb to 837,000 lb and payload by 53,000 lb. The incidental impact of AAR on C-5B is to raise the airborne payload capacity by 83,000 lb: the difference between the weight the aircraft can lift off the ground and the weight it can sustain when cruising.

The Russian equivalent of C-5 is the AN-124 Condor, which made its Western debut at the Paris Air Show in 1985. As with most Soviet aircraft, technical data on it is derived from selective official releases and informed guesswork by Western aviation specialists. Like the Galaxy, Condor has a visor-type forward-loading door and an undercarriage which can lower and raise the aircraft to facilitate ramp loading. Both aircraft are equipped to carry pallet-mounted loads; in the C-5 they are moved by a hydraulically operated winch; in the AN-124 by four electric cranes which travel on rails in the freight hold roof. Both are designed to operate from hard-packed airfields with rough taxiways. But whereas the C-5B is essentially a modified C-5A, which itself was a product of 1960s aircraft technology, AN-124 was observed to have benefited from advanced wing design, a four-channel analogue fly-by-wire system, the incorporation of glass fibre-reinforced plastic and carbon fibre composite sections in the fuselage, and the use of titanium alloy planking for the cargo floor. All these measures together will have reduced proportional weight by several thousand pounds and contributed to a reported theoretical payload advantage over C-5 of 25,000 lb over comparable

PLATE 7.8. Strategic airlift: USAF Lockheed C-5B Galaxy.

distances. Its four Lotarev D-18T turbofan engines generate 51,650 lb of static thrust each, considerably more than the uprated engines of the C-5B which had been fitted with four General Electric TF39-GE-1C engines of 41,000 lb of basic thrust each, but all capable of temporary augmentation to 43,000 lb for heavy load take-off. The fuel consumption of the Soviet aircraft was, however, not known and it is that factor, rather than the raw power of the engines, which strongly influences range/payload characteristics. As a result, by early 1986 no consensus existed on the relative gross range/payloads of the two strategic transports.

There was, however, a great deal of information available about another transport aircraft designed to enter USAF service in the early 1990s. The McDonnell Douglas C-17 (see Plate 7.9) is to meet a requirement to move heavy and bulky loads between and within theatres of operations, into and out of 'austere' landing fields. 'Austere' covers a wide range of contingencies, but could include airfields with limited approach and landing aids, short and compacted unpaved runways; minimal or no refuelling and maintenance facilities; and little or no external loading and unloading equipment. The promised performance figures for C-17 illustrate very clearly the combined impact of several state of the art military aviation technologies.

The aircraft is powered by four Pratt and Whitney PW-2037 turbofans each producing 37,000 lb thrust. They are fitted with directed flow thrust reversers which can be deployed in flight as well as on the ground, where they can back a fully loaded aircraft up a 2.5 per cent incline and greatly facilitate manoeuvring in restricted spaces. To reduce the likelihood of danger to the engines from ingestion

PLATE 7.9. Artist's impression of McDonnell Douglas C-17 advanced cargo aircraft.

of stones and other foreign bodies encountered on unpaved surfaces, the exhaust from the thrust receiver is deflected upwards. Like the AN-124, C-17 has a 'supercritical' wing, with a flat top and undercut trailing edge swept to 25° which gives greater span for a given weight, producing greater aerodynamic efficiency and coincidentally greater capacity for fuel tanks. Final approach and landing speeds will be greatly reduced because of an artificial airflow generated from the engines to double-slotted flaps, thereby generating increased lift: of great importance in short field capability. The flaps themselves will be made of titanium which, together with some 9,000 lb of different composite materials, combine to effect a major reduction in empty aircraft weight.

Avionics technology developed for McDonnell Douglas combat aircraft has been incorporated in the C-17 together with that associated with contemporary civil airliners.

The fit will include digital autopilot, colour weather radar, automatic air-to-air location systems for rendezvous with tankers, station-keeping instruments for night and all-weather formation or stream operations essential for concentrated paratroop dropping, dual inertial systems and a flight management computer. Information, including instrument serviceability states, emergency procedures and routine check lists, will be presented on four CRT displays and two head-up

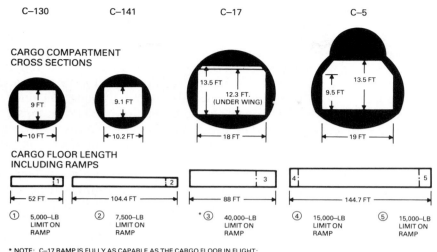

FIG. 7.3. Transport aircraft freight capacity.

displays (HUDs). The HUD will facilitate precise touch-down and contribute still further to short field performance.

Freight capacity (see Fig. 7.3) has been maximised by integral design as well as weight reductions elsewhere. The C-17's freight hold is only 1 ft narrower than that of the C-5 and the same height, although some 60 ft shorter. It will deliver 172,000 lb over 2,000 miles to an austere airfield and still have enough fuel to fly on to a further base, thereby obviating the need to transfer loads from a 'strategic' to a 'tactical' transport. C-17 will carry the loads of two C-141 Starlifters and can drop three 18.5-tonne infantry fighting vehicles by low-level parachute extraction, or paratroops and lighter loads at heights up to 25,000 ft. The entire aircraft is designed to be operated by a crew of three: pilot, co-pilot and loadmaster, reflecting the perceived need in the West to reduce manpower requirements as much as possible. Economy is completed by a considerable reduction in the need and frequency of ground maintenance.

Long Range Navigation

The C-17 is, like most other military freighters, a hybrid, in that they may spend most of their operational life discharging military tasks in peacetime flying on established civil airliner routes. On the other hand, they could be used in clandestine operations in times of tension and in war when, again like surface, shipping, they would not wish to advertise their position to an enemy. Consequently, C-5, C-17 and AN-124 carry some navigation aids which can be used with civilian ground-based transmissions, and others which cannot. C-5, for example, has two VOR/ILS Marker Beacon systems for bad weather approach and landings; two TACAN systems for radio navigation; two Collins AN/ARC-186 VHF/AM/FM radios with both clear and secure modes; two HF single side-band radios; three inertial navigation systems; a Collins DF-206 low-frequency automatic direction-finder; and a Bendix colour/weather/mapping radar. The whole is coordinated by a digital Standard Central Air Data computer and is monitored by a further computer (MADAR II) to detect systems malfunction. AN-124 is similarly, but not identically, equipped; with three INS, two radars for weather and mapping, Loran and Omega radio position-locating systems, and in case all systems should fail, an old fashioned astrodome. Although the pilot and co-pilots' instruments are conventional electromechanical, AN-124 is apparently also equipped with a computerised monitoring and fault-diagnosis system which shares a CRT display with the computerised cargo-loading and monitoring system.

It is probable that in the near future not only transport aircraft but those more directly involved in offensive and defensive operations will be able to navigate much more accurately as a result of satellite-derived navigational systems. In the West, the Navstar Global Positioning System (GPS) will comprise eighteen satellites uniformly spaced every 120° in six orbits 10,898 nautical miles above the Earth and at an inclination of 55° to the equator. Each will take 12 hours to complete an earth orbit and all will be monitored from the ground to ensure maintenance of correct coordinates. Each satellite carries four atomic clocks which would take 300,000 years to accumulate a 1-sec error. A microprocessor stores navigational data fed from ground control and each satellite transmits a uniquely coded spread spectrum multiple signal. Three frequencies transmit to civilian and military receivers, but with different codes. Transmission times from satellite to receiver are divided by the speed of light, giving the distance between them, while measurement of the Doppler shift in the received signal indicates how the satellite's position is changing. By monitoring four satellites at the same time, the military receiver will receive a three-dimensional fix on position to within 10–20 meters anywhere in the world; its velocity to 0.1 miles/sec and exact time in nanoseconds. The satellite transmissions are difficult to jam and are probably hardened against EMP. Collins, the manufacturers, emphasise the military significance of such accurate navigational and velocity information to update INS systems in strategic aircraft and in long-range missiles requiring midcourse or even continuous updates and guidance. In the case of missiles, the technology would be complex and expensive, but the prospect, however theoretical, of guiding air-breathing vehicles by satellite signal either in addition to or instead of from an aircraft would extend considerably options for offensive and defensive operations.

Frequently, however, there is an Achilles heel somewhere in emerging technology. In the GPS it may be in the ground control segment which is located at Vandenberg AFB in California with outstations in Alaska, Guam and Hawaii. Each station transfers data on the satellites to Vandenburg from where corrective action, if required for the satellites, is computed for upward transmission. Nevertheless, Navstar GPS offers an increase in navigational accuracy together with velocity computation which has dramatic significance for all air operations, but especially for those whose effectiveness is heavily determined by exact positioning and velocity of the airborne platform or missile.

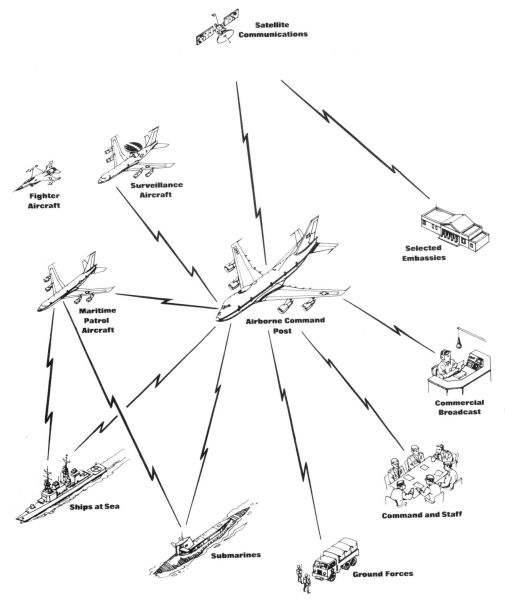

FIG. 7.4. Strategic airborne command and control.

E-4 AIRBORNE COMMAND POST

Should deterrence ever fail, and the superpowers engage in a conflict likely to engulf large parts of the world, strategic air power would assume a global responsibility hitherto held only in reserve. Because of the accuracy and

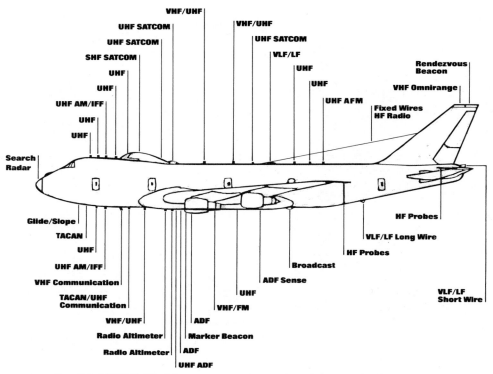

FIG. 7.5. USAF E-4B airborne command post aircraft communication systems.

destructive power of aerial weapons already described, quite apart from those delivered from land and sea, fixed, static installations of every kind are increasingly vulnerable. Sustained command and control of armed forces in any conflict is essential; in the maintenance of a credible deterrence posture, the perceived ability to retain such control in the face of nuclear threat is paramount. As a result, the United States has elected to construct an alternative airborne command post, linking national government, military authorities and retaliatory forces world-wide (see Fig. 7.4).

In 1974, the first of three modified Boeing 747-200 airliners entered USAF service as the E-4A, equipped largely with communication systems hitherto in use on earlier EC-135s. In 1979, a new 747–200 configuration, the E-4B, entered service and by 1985 all three original E-4As had been modified to E-4B standards. The EC-135, similar in size to the Boeing 707 series, had a usable floor space of 873 ft^2, E-4B has 4,620, and carries three times the earlier payload for a refuelled endurance of 72 hours with up to ninety-four crew members. Aircraft operation and maintenance are self-contained, allowing operation from and recovery to a very large number of civilian and military airfields, thereby further reducing the command post's vulnerability to surprise attack. The communications links on E-4B illustrate the extent and complexity afforded to airborne installations by modern signals technology (see Fig. 7.5). A 1,200 kVA generator produces power for thirteen external communication systems operating through forty-six antennae. In addition to a full suite of navigational aids, which will presumably include Navstar GPS, E-4B has both super-high-frequency (SHF) and ultra-high-frequency (UHF) satellite links which reduce dependence on ground stations and reduce vulnerability to jamming and detection by direct tracking. A 5-mile-long trailing wire antenna provides very-low-frequency and low-frequency communication with submerged submarines, resistant to atmospheric nuclear effects and jamming. Additional two-way radio links are provided on HF, MF, VHF and UHF bands, while the whole system could be linked with commercial telephone and radio networks to relay broadcasts to the general population. All systems are hardened for protection against the electromagnetic effects of a nuclear burst and are integrated with a microcomputerised automatic digital network. The considerable space of the 747 airframe is utilised for a National Command Authority area, conference room, briefing room, battle staff work area, communication control centre, technical control monitoring centre, crew rest area, on-board maintenance area, winch operator's station for the VHF antenna and extensive electronic equipment.

In sum, at one end of the air power spectrum is the combat aircraft and its missiles, influencing the course of a conflict in its own immediate environment. At the other, is the familiar outline of the jumbo jet, but now distinguished by the outward manifestations of its internal military technology which has added a further, indeterminable dimension to the concept of 'strategic' air power.

THE AGGREGATE

In the history of warfare, eighty years is a minute interlude. But in that period the impact of technology has brought air power from the initial role of marginal player in the land and sea battle to the present, when it dominates every dimension of warfare and could, in the direst circumstances, provide the only means to wage any kind of warfare at all. Perhaps for the first time since the dreams of the early air power theorists, air power will be constrained not by the limits of technology, but by the limits on man's own intellectual capacity to master it.

Index